Walter J Malden

Sheep Raising and Shepherding

A Handbook of Sheep Farming

Walter J Malden

Sheep Raising and Shepherding
A Handbook of Sheep Farming

ISBN/EAN: 9783742860811

Manufactured in Europe, USA, Canada, Australia, Japa

Cover: Foto ©berggeist007 / pixelio.de

Manufactured and distributed by brebook publishing software
(www.brebook.com)

Walter J Malden

Sheep Raising and Shepherding

SHEEP RAISING AND SHEPHERDING:

A HANDBOOK OF SHEEP FARMING.

BY

WALTER J. MALDEN,

Principal of the Agricultural College, Uckfield;

*Late Resident Supt. of the Royal Agricultural Society's Experimental Farm,
Woburn; Late Professor of Agriculture, Downton Agricultural
College; Late Agricultural Supt. of the Royal
Dublin Society;*

AUTHOR OF "THE POTATO IN FIELD AND GARDEN," "RATIONAL PIG-KEEPING,"
"TILLAGE AND IMPLEMENTS," "FARM LABOURER'S TECHNICAL
INSTRUCTOR," "CROSS-BREEDING IN SHEEP," "THE
CLUN FOREST SHEEP," ETC.;

ASSOCIATE OF THE SURVEYORS' INSTITUTION.

ILLUSTRATED.

LONDON:

L. UPCOTT GILL, 170, STRAND, W.C.

1899

PREFACE.

THE farmer's "sheet anchor," as the sheep has so aptly been termed, during the past twenty years of dark days in British farming, in spite of a greatly extended foreign competition, still remains, and is likely to remain, the key to successful farming over the greater part of the country. Trustworthy books describing the management of sheep in Scotland have been plentiful, but those dealing thoroughly with English sheep management are few, therefore the greater part of this book is given up to English methods; though, as the broad principles of sheep management hold good throughout the world, it has its bearing on Scottish sheep also—the success of the Shropshire Downs in Scotland being a case in point.

Brought up on a farm in the South-East Midlands where 700 to 800 sheep were wintered yearly, about 300 of which were bred on the place, my earliest recollections were of the Leicester and the Cotswold at a time when they were giving way in that district to the Oxford Down, and more recently to a cross with the Oxford and Hampshire Down. Leaving there for Wiltshire, four or five years were spent among the Hampshire Downs, to be followed by a short residence among the Shropshire Downs, and lately, among the Southdowns and Kentish; which, together with an occupation that has

taken me to all parts of England and Ireland, has given
exceptional opportunities of following the methods of
management of the several breeds of sheep, and of comparing the peculiar features of the breeds.

This book is thus a record of a personal experience, and
if it differs in any respect from that of individuals, I
trust it will be accepted as it is proffered, as a guide
to those less experienced, and as a summary of the experience gained with the sheep and by intercourse with those
who have done so much to' improve the more notable
breeds. In calling to mind many of these, I always have
prominently before me a kind and helping friend, during
many years, in the late Mr. Charles Howard, of Bidden-
ham, than whom no one did more to build up the breed
of Oxford Downs, and to whom, when before his death I
began to write these pages, I had hoped to dedicate them,
as he first gave me an interest in the breeds of sheep.

Every breed and every district possesses distinct features
requiring different methods of treatment, and in the following pages I have endeavoured to deal with them
broadly, at the same time entering as fully as possible
into the details of management as practised by the best
authorities.

Other textbooks deal ably with the history and characteristic features of the breeds of sheep, and it has not
been thought necessary to devote much space to them here,
but rather to enter fully into the details of shepherding.

The excellent illustrations of breeds which week by
week are given in the leading agricultural journals, do
away with the necessity for showing all in a book on
Shepherding: some of those given here are meant to illustrate from what skilled breeders have succeeded in producing
the sheep of to-day—a work mainly carried out within this

century, and in no small degree during the latter half of it.
That they have followed sound laws of breeding, and at
the same time kept in view the requirements of the markets and the changes which have from time to time been
made in the systems of farming, is obvious. Moreover, the
soundness of the principles they have adopted is further
shown by the fact that every sheep-breeding country, no
matter how much differing in soil and climate, finds it
necessary to come to Britain for fresh blood to maintain
its flocks.

It is believed that the divisions made in the book,
particularly those showing the daily management of the
ordinary sections of the flock, as well as the management
from one portion of the year to another, will be helpful
to those who are commencing their experience in Shepherding.

W. J. MALDEN.

AGRICULTURAL COLLEGE AND TRAINING FARM,
  UCKFIELD, SUSSEX.

CONTENTS.

SHEEP RAISING AND SHEPHERDING.

CHAPTER I.

INTRODUCTORY.

A GOOD sheep farmer is generally a good all-round farmer, for no class of farm stock requires closer attention or more skilful management than sheep. Moreover he must be a skilled tiller of the land, as the ability to secure suitable crops at all seasons proves he has his arable land well under control. When these are ensured, sheep farming is generally one of the most profitable sections of the farm.

The sheep farmer must always look well ahead to provide crops for all divisions of his flock; and in unfavourable seasons must be resourceful, so that if some portions of the food he might reasonably rely on prove to be failures, he may devise means which will enable him to carry his flock through a critical period. Such occasions occur frequently, and if the farmer has made no special provision, the sheep have to be sold in a bad market, and subsequently he has to buy in again on unfavourable terms.

As a rule, the sheep management throughout a district varies very little in broad principles, although there is often considerable variation in the details. What appear to be minor details frequently make all the difference

B

between profitable and unprofitable sheep keeping. The customs of districts are generally sound, as they are the outcome of a long experience, which has shown what is best under the particular conditions of soil and climate; it is, therefore, unwise to make profound changes unless all features of the changes are well considered previously, and are proved, at any rate on a small scale, to be more advantageous than those practices they are to usurp.

Without change, however, improvements cannot be made, but alterations are not likely to prove valuable unless they are made by those who possess exceptional skill and experience. Those with great knowledge of the markets, and who are able to judge from the amount of food there is likely to be throughout the country at a given time, are able to foretell whether prices are likely to be high or low at a particular date in the future. These are justified in consuming their fodder at the customary period, or may reserve it so as to have it available at a time when sheep keep is likely to be scarce. For instance, in dry summers it is often difficult to get a good plant of turnips; and even transplanted crops, such as kale and cabbages, which, as a rule, may be successfully planted at a much later date than is possible for swedes and mangolds to be sown to produce a remunerative crop, do not thrive. When this happens, the spring sheep feed throughout the country is necessarily small in quantity. Yet the same number of sheep have to be supported as in a more plentiful year. To keep the sheep healthy and thriving, the supply of spring food has to be drawn on, leaving little for that period known as " between roots and grass," which is usually from March to May, and depends very much in duration on the earliness of spring growth.

In seasons of great shortness it is often particularly

remunerative to save the food until spring, and then to buy in sheep, which, as others cannot afford to keep them, are necessarily sold at reduced prices; but when the fresh supply of summer keep becomes available they rise in value again, thus giving to persons in a position to purchase them in times of scarcity of food the advantage of the rise in price, in addition to the increased value due to their greater size and improved condition. When food is thus saved until spring, a much larger number of sheep can obviously be kept during the short period mentioned than if they had been kept throughout the winter, yet each individual sheep will probably pay more for the short time than those which were bought in the autumn. It is to an extent a speculation: so, however, is the purchasing of sheep in the autumn, when the risk of loss of life and disease has to be run for a long period. Many thousands of sheep are bought in in the autumn in such years, and the buyers know there is practically no chance of profit; yet, because it is the custom of the district to buy at that season, they do it year after year. In no other business would such a course be followed.

The time to buy is when there is a fair chance of profit, and not when there is little likelihood of gain. The skill of the farmer comes in in buying at the right time; he must watch the markets closely, and buy in accordance with the trade and the prospect of an early return of the time when keep will again be plentiful. If he misses his chance by being over-greedy, he may miss much of the profit he otherwise had it in his power to obtain.

Late spring feeding of roots and other fodder crops tests the skill and resourcefulness of the farmer in the manage-

ment of his land and cropping, as he throws an increased
amount of work on his horses late in the season, and it
will probably necessitate an extensive alteration in the
crops he must sow. This is easier now than it was a
few years ago, as few restrictions are made as to the
order of cropping; whereas when narrow leases restricted
farmers to a strict rotation it was almost impracticable.
It will therefore be seen that only those possessing special
skill are justified in going outside the ordinary practices
of the district.

Other illustrations might be given as to the stocking
of the land in accordance with the prospect of ample or
deficient keep at a particular future date, and of the
consequent variation in prices, which allow the shrewd
farmer an opportunity of obtaining an exceptional profit.
They are by no means rare; but the one given should be
sufficient to illustrate the manner in which a shrewd man
may obtain remuneration for the exercise of his greate.·
skill; and also to indicate to those less experienced that
the subject of sheep management can only prove entirely
remunerative when everything bearing upon it is thoroug'ly
mastered.

The management of the food when grown so as to make
the best of it, and so that the land may not be injured
whilst it is being consumed, and the management of the
sheep themselves, are dependent on the skill of the farmer
backed up by the skill of the shepherd.

CHEVIOT EWES.

Bred by Mr. Robson, of the Byrness.

CHAPTER II.

THE BREEDS OF SHEEP.

Classification—The Making of Breeds—Selection of Breeds—Soil and Climate—Cross-breeds.

IT is not intended to devote much space to the description of breeds, as the purpose of this book is to deal with the management of the animals with a view to profit. At the same time it is convenient to give a brief description of the more important varieties, as it will save a considerable amount of repetition in future pages, and may be useful for reference.

Classification.

The breeds of sheep for the practical purposes of the farm may be divided into those which are essentially (1) sheep of the rich grass land; (2) sheep of the arable land; and (3) sheep of the hills, heath, and other poorly-clad grass land. These are to a great extent distinguished by the wool, the rich grass land carrying as indigenous (if improved and somewhat alter. in character) breeds with long wool and white faces; arable land, those which generally originated on down land, possessing short wool and black faces; and hills, those with coarse, harsh wool varying in length, and with black or mottled faces, though this does not always hold good, as evidenced by such breeds as the Welsh Mountain and the Cheviot, which have soft short wool and white faces.

Another division which holds fairly good is that of long-wools, medium-wools, and short-wools.

The table given on page 7 shows the divisions of the more notable breeds, conveniently arranged. A few ·other breeds are essentially local, and are not of general interest.

The short-wool sheep are prized chiefly for the excellence of their mutton, although the wool is fine, and is valuable for special purposes. The medium-wool sheep produce good mutton with a fair fleece. The long-wools are not as a rule much prized for their flesh, as it is coarse and fat in comparison with the Down and hill sheep ; but the weight of the wool and carcase and their suitability for rich pasturage render them profitable on the land where they are met with.

Few of these breeds are of pure descent from the indigenous stocks of the country with which they are identified. Before the enclosure of common lands, and when the sheep moved over great heaths and in forests, the breeds rarely intermixed ; but when they were brought into enclosures the breeding was regulated by the farmer, who mated them according to his fancy. Two breeds have had the greatest share in the making of modern breeds. The long-wool breeds have been improved by the Leicester, and the short-wool breeds by the Southdown. The Leicester was taken in hand by Bakewell about the middle of the eighteenth century, and by careful selection he wrought great improvement in the breed, and built up for it a big reputation. Its renown spread, and breeders from all parts of the country purchased rams to mate with their own sheep. The best results were obtained on long-wool breeds ; in fact, it is doubtful if any great permanent good was effected by

Breed.	Wool Long, Medium, or Short.	Colour of Face.	Habitat, or conditions under which they are most often kept.	Horns.
Southdown . .	Short . .	Grey black, not slaty or sooty	Downs, grass and arable .	None
Hampshire Down .	Short . .	Black to mouse colour .	Downs, but chiefly arable	None
Shropshire Down .	Short . .	Black	Good grass and arable .	None
Oxford Down . .	Medium .	Blackish, not speckled .	Chiefly arable . .	None
Suffolk Down . .	Short . .	Black	Chiefly arable and heath .	None
Clun Forest . .	Short . .	Mottled tan and white .	Hill and slightly arable .	Occasionally
Kerry Hill . .	Short . .	Black and white speckled .	Hill and slightly arable .	Occasionally
Cheviot . . .	Short . .	White	Hill and grass . .	Occasionally
Welsh Mountain .	Short . .	White, sometimes tan .	Hill	Frequently
Blackfaced Scotch .	Long, hairy .	Black and mottled . .	Grass and hill . .	Strong
Dorset . . .	Medium .	White	Grass	Strong
Wicklow (Irish) .	Medium .	White	Grass and hill . .	None
Kerry (Irish) . .	Short . .	White	Hill	Slight
Devon . . .	Short . .	White	Grass	None
South Devon . .	Long . .	White	Grass	None
Dartmoor . .	Medium—long .	Whitish . . .	Moor and grass . .	None
Exmoor . . .	Medium—long .	Whitish . . .	Moor and grass . .	Slight
Herdwick . . .	Medium .	White with brown specks .	Hill and grass . .	Curly
Lonks . . .	Long, hairy .	Black and white . .	Hill and grass . .	Rams generally horned
Cotswold . .	Long . .	White to grey . .	Grass	Short
Leicester . .	Long . .	White, rare spots . .	Grass	None
Border Leicester .	Long . .	White	Grass	None
Lincoln . . .	Long . .	White	Grass ⎫	None
Roscommon (Irish) .	Long . .	White	Grass ⎬ and to some ex-tent arable.	None
Kentish or Romney Marsh .	Long . .	White	Grass ⎭	None
Wensleydale . .	Long . .	Bluish . . .	Grass	None
Ryeland . . .	Medium .	White	Grass	None

7

it on any breed of short-wool sheep. On these latter, however, the Southdown had a most beneficial effect, and all Down breeds have been improved by them since John Ellman of Glynde took on himself to improve the Southdown, towards the end of last century. The Leicester improved the quality of the wool of other long-wool breeds, and imparted finer quality and earlier maturing properties to them. The Southdown improved the quality of - the meat and wool, toued down the conrseness of the bone, made the sheep more compact, and brought them to early maturity, without impairing their vigour. As the Leicester effected little good on the short-wools, so the Southdown left but few good impressions on the long-wools.

The improvement of every breed now recognised has steadily advanced throughout the century. At first a great deal was done by means of crossing with other breeds, so that the valuable characteristics of the breeds employed might be imparted to the offspring. Latterly the improvement has been more particularly brought about by selection, improved methods of management, and the introduction of new systems of farming, although even now a limited amount of crossing is done. In some breeds, especially those in which the Leicester had an important share in the development, it is necessary to occasionally infuse a fresh strain of Leicester blood to keep them up to the type required.

The Making of Breeds.

The Southdowns and Leicesters have been mentioned as having been the great improvers of the breeds found in the country at the present time. As a point of interest, as examples of what has been done to develop the breeds within the past century, and to encourage others to carry on the work, a few of the breeds as they existed in the early part of the century (see Plates), as well as some of the animals as we now find them, are illustrated here. The illustrations of the older types are taken from the collection of paintings from life executed for the Agricultural Museum of the University of Edinburgh by the late Mr. W. Nicholson, R.S.A.; and the animals were selected from the flocks of the best known breeders of the day.

The " old " Southdown ram (see Plate) was from the flock of Mr. John Ellman, of Glynde, and was an animal far more improved than is either of the other Heath breeds illustrated, and it will be realized how such an improved animal would impart good features to them. When, however, the Southdowns shown in the group forming the Frontispiece of this book are compared with this sheep of sixty years ago, it will be seen that a great advance has been made towards early maturity. In Chapter III. the points of a sheep are dealt with, and in comparing the sheep of *then* and *now*, it may be well for those who have not had much experience with sheep to refer to those points. If this is done, there is little need for minute description or comparison here.

Hampshire Down breeders will see little in the Old Wiltshire sheep, bred by Mr. Turner, of Hindon, Wiltshire, (see Plate) to remind them of their modern animals. The most striking feature is the Roman nose, which is noticeable in the Hampshire yet. The Hampshire Down (Fig. 1 and Plate), however, is the offspring of the Old Wiltshire and the Southdown. The Old Berkshire Knot was also crossed with the Southdown, and probably there was a merging of the three breeds in some cases ; but it was the two first-mentioned breeds which mainly built up the Hampshire Down. The Old Wiltshire had several good features, but were very slow in maturing—vastly different from the modern sheep, which provides so large a proportion of the fat Easter lamb. The recurrence of horns in modern sheep not carefully looked after can easily be understood when the illustration of the old breed is observed; and the

Fig. 1.—Hampshire Down Ram.

desirability for not breeding from those showing rudimentary horns is obvious.

The Plate showing the Old Norfolk breed is particularly interesting, as under the name of the Suffolk breed it is the breed of Heath sheep last taken up in a systematic manner for improvement; and it has made its greatest progress within comparatively few years. There are, of course, old flocks to which careful attention has been given for a long period, but it is in point of improvement the youngest pure breed. Within the past quarter of a century it has formed one of the best object lessons in breeding. The modern animals (Figs. 2 and 3) show great variation from the horned animal seen in the Plate. The long legs, narrow body, light thighs, and other features of

Fig. 2.—Suffolk Shearling Wether.

an inferior animal have given way to sheep possessing the characteristics of early maturity. The lamb beside the ewe in the Plate affords direct evidence of the crossing of the Norfolk sheep with the Southdown, for its sire was a Southdown.

The sensational sale in July, 1898, of the Lincoln ram belonging to Mr. Dudding (see Plate), when for the first

time in England a sheep realized 1,000 guineas, calls for
special attention. Bakewell found the Leicester in an unimproved condition, and developed it into an animal of finer
quality in every respect, calling his improved animal the
New Leicester (see Plate). It was this breed which was instru-
mental in improving all the long-wool breeds. It is noticeable
in the Plate that the signs of good breeding are more marked

Fig. 8.—Suffolk Ewe Lambs (5 months).

than in those of any other breed of the time. The great
heart-measurement, well-set-on tail, full brisket, well-set legs
and kindly character generally are indicative of a sheep's
high breeding. Another Plate shows the Old Lincoln sheep,
a breed of great size, but slow to mature. The flanks
were hollow and the sides flat. It had, however, a fleece
of long unctuous wool which the mating with the Leicester

further improved. The Plate of the 1,000 guinea shearling ram shows what fine character has been imparted to the breed; and it is greatly to the advantage of the breeder that it mates so well with the Merino, causing great demand for it in the great wool-producing countries of the Southern Hemisphere.

The illustrations of old breeds given as Plates are sufficient to show the lines on which modern breeds have been developed. It does not require a very great stretch of imagination to realize how other old breeds not shown here have taken their modern form. Next in importance to the early maturity now obtainable is the fact that by following sound lines of breeding the constitutions of the animals have not been seriously impaired. A safeguard against the undue weakening of the animal is that when inbreeding is too persistently followed evil results quickly show, so that it is known among all breeders that it must be avoided. There are instances where too close breeding has produced sheep which showed signs of insanity.

Selection of Breeds.

The selection of a breed suitable for a particular locality is dependent on many conditions. The first and most important is the soil, after which come the climate, system of farming, nature of herbage, and purpose for which they are required. Down sheep are not best suited for rich pastures, and long-wool sheep are not profitable on Down land. To a less degree all other sheep are influenced by the soil on which they are carried, the extent being regulated by the dissimilarity of the soil to that to which they are indigenous.

A very successful exhibitor of Shropshire sheep in Nottinghamshire divided his flock, one portion being kept at home on a good loam, while the other was sent but a few miles away to a sandy soil; the result was that he continued to win prizes on the loam, but on the sand the flock lost size and characteristic features, and so deteriorated, notwithstanding every effort to prevent it, that in a few years he was obliged to substitute another breed for them.

Hampshire Downs taken on to gravels and cold loams quickly lose their type, and are not so profitable as on their native chalks, although the sheep may be moved but a few miles. Much of this is due to the colder lair in winter, as in summer time they do well. With other breeds the same variation in type is experienced when they are removed from conditions under which the breeds were built up. It is therefore important to regard the nature of the soil when entertaining the idea of making a change.

There is, however, one important point to be remembered in the selection of a suitable breed for a particular farm. Although the conditions may not be favourable for breeding, they may be for feeding. When breeding, it is usual to keep to a selection of the breed for a number of years, during which the sheep gradually change their type; whereas during the few months when imported sheep are being fed little change takes place, and if it does, it need not be of great moment. As a matter of fact, the greater portion of the Down-bred sheep are fattened off on land at a considerable distance from their breeding place, and thrive well.

Soil and Climate.

Climate, undoubtedly, has an effect on the thriving capabilities of sheep. Those accustomed to dry lair, a moderate rainfall and mild climate, suffer when the conditions are changed. It is noticeable that the indigenous breeds of the southern portion of the country, where the rainfall is light, the climate mild, and the lair is dry, are almost without exception short-wool breeds. An exception is found in the Romney Marsh sheep, but exceptional conditions account for this: they are indigenous to the rich low-lying marsh land of a small tract in the dry climate of Kent. The proximity of the sea, and the low-lying position of the land, however, counteract other influences, and a long-wool breed is found. Dryness of the climate has an effect on the herbage beyond that which is caused by the nature of the soil. Down land naturally carries short herbage, but the herbage of the Downs differs from that of the thin soils of the limestone situated in more northern and wetter parts, where a different type of sheep is carried.

A great effect of moisture on land is that sheep which are indigenous to it or have been carried on it for a long period open their claws or digits when they tread on it. This is necessary to prevent them from sinking deeply, whereas on dry soils the feet open very little. Thus, on the Downs, where the ground is rarely sufficiently moistened to allow the sheep to tread through the turf, the feet are small and the digits close, as compared with those of sheep carried on loose soils supporting rich pastures.

The formation of the feet has an important bearing on their liability to foot-rot. Sheep which for generations have had to expand their feet have developed a hard skin between the digits, and this is not easily abraded when it becomes moistened, as when brought into contact with long wet herbage, or when the sheep are penned on wet gritty arable land; whereas under similar circumstances the close-toed sheep are very susceptible to fracture of the skin. When once the skin is broken, the germs of the foot-rot disease easily establish themselves, and the foot becomes diseased. A difficulty is, therefore, experienced in placing Down sheep on soils carrying rich pasturage, or on arable soils which produce friction about the feet.

The suitability of the sheep for the purpose which is likely to prove most remunerative is an important matter. Generally, in these days of foreign competition, meat is the first object, as wool is imported at such a low price; still, the value of a fleece which weighs ten or more pounds is not to be ignored. The type of sheep must be decided to no small extent by the characteristics of the farm and the kind of cropping it will carry with most success. It is fortunate that while mutton from white-faced sheep is not so valuable per pound, the sheep produce large carcasses and abundant fleeces of wool, which together are of great value; so the sheep-grazier is not altogether outplaced by the arable land farmer, who, owing to the greater expenses to which he is put in obtaining crops, requires some advantages to make matters equal.

Cross-Breeds.

As we recently discussed the subject of cross-bred sheep in the *Journal of the Royal Agricultural Society* (Third Series, Vol. VI., Part II., 1895), which is within the reach of most readers of this book, we will deal but briefly with it here.

Pure breeds of sheep—that is to say, those which are derived from one stock only—are few in number, the most important being the Leicester and the Southdown. Almost all other breeds as at present found are more or

less crossed, as the Shropshire (Fig. 4 and Plate), which
was derived from the Morfe Common and the Southdown;
the Hampshire Down, from the old Wiltshire horned sheep,
the Berkshire Knot, and the Southdown; the Oxford (Fig.
5), from the Hampshire Down and the Cotswold; the
Lincoln, from the old Lincoln and Leicester, and so on.

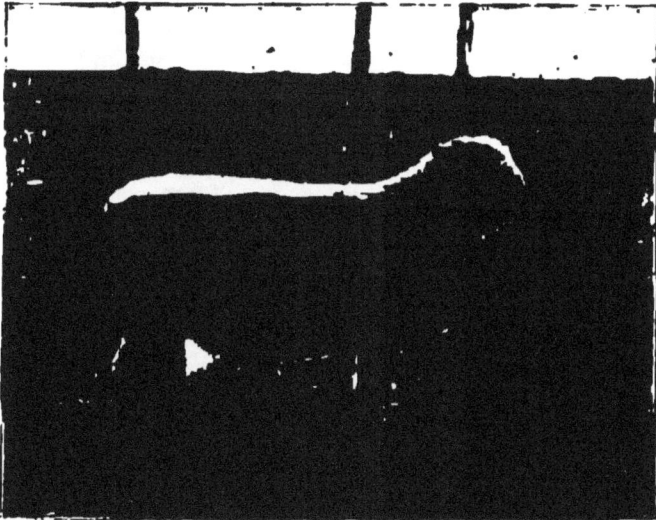

Fig. 4.—SHROPSHIRE SHEARLING RAM.

The hill breeds have been less systematically crossed. The
three Cheviot ewes shown in the Plate illustrate the trueness
with which the breed has been bred. In the time of Bakewell this breed was crossed with the Leicester, producing
the Border Leicester. They have long been quite distinct
breeds. The amalgamation of the hill and plain sheep forming the Border Leicester, has produced a breed peculiarly

C

suitable to districts both north and south of the Cheviot Hills. The greatest amount of crossing has occurred where the local agriculture has altered most, as by changing the system of cropping fresh kinds of feed have been introduced, permitting a different class of sheep to be carried.

Where, however, a particular breed holds a specially

Fig. 5.—Oxford Down Shearling.

prominent position in a large district it is generally found that it originated from a breed which was indigenous to the district, although it may have been considerably altered by the introduction of some one or more varieties foreign to the district. An exception is found in the case of the Oxford Down. This breed is most commonly met with on loamy soils in districts where arable farming largely prevails, and where the indigenous breeds have been exterminated. The altered conditions of farming demanded a composite breed, which would thrive on those soils under the changed circumstances. Long-wool sheep do not produce meat of sufficiently good quality, and the strictly Down breeds find these soils too cold, and are liable to succumb too readily to foot-rot.

CHAPTER III.

THE POINTS AND NOMENCLATURE OF SHEEP.

Points—Handling—Dentition—Nomenclature.

Points.

BEFORE dealing with the management of sheep, it is advisable to discuss some of the points which should be looked for in them, as badly-bred sheep are rarely profitable, or, at any rate, are not so profitable as those better bred.

In the first place, a sheep should possess features and an outline which are pleasing to the eye. The animal should have a well-balanced appearance, otherwise there is some feature lacking or too prominent. A lean sheep can never show to such great advantage as one which is fat, as meat fills up the frame, and makes it more even and level. It is not, however, necessary that an animal should be fat for its good points and thriving properties to be noticeable. There are signs of good breeding, rapid maturing, good wool, and growthiness in a well-bred sheep, no matter how poor it may be or how ragged the wool may appear, and these are readily seen by a good judge of animals. Features which are good in one breed are generally good in another, though, of course, there are characteristic features in every breed distinguishing them from other breeds; but in broad principles that which is good in one must be looked for in another.

The *body* should have a well - squared appearance; neither end should taper; the hind quarters and the fore quarters should finish boldly, and the line of the back and that of the belly should be parallel.

The *tail* should be well set on in a line with the back, so as to give the appearance of finish. It should not be too high, and if too low it denotes slackness of the hind quarters and coarse breeding. The tail, or dock, should be broad, affording a good grip when taken in the hand.

The *loin* should be broad and flat. The sheep should "back" well. If the spine rises high about the hind quarters the sheep is unthrifty, as it is in an "unimproved" condition. It will generally be seen that where the backbone is raised high on the loin and rump, the girth through the heart is small and the fore quarters are light. These are conditions which, almost invariably, are present in wild sheep and those which have not been subjected to improvement by careful selection and breeding. Almost all old illustrations of sheep show these characteristics. Broad, deep *fore quarters*, and broad, flat loins indicate powers of early maturity, consequently they must be looked for before any other features. To those accustomed to flat-backed, white-faced sheep which carry a large amount of fat on the back, Down sheep appear to be narrow and high in the loin, as they do not lay on so much fat in that part unless specially fed. In comparing the touch of a long-wool sheep with a Down, a long-wool on which the backbone appears to be at all prominent is not ripe, whereas if the same amount is felt on a Down sheep it may still be in good condition for the butcher. Those accustomed to handling Down sheep must therefore be careful when handling long-wools, or the condition of the latter may be over-estimated.

The *neck*, or *scrag*, should be broad, as a thin neck is

usually associated with a narrow fore quarter. The neck should taper fully from the body and shoulder, and be brought up from a square, deep brisket. Where this is the case the sheep "meets one" well. The *brisket* should show distinctly in front of the fore legs when viewed from the side. The neck and head should make a bold, level sweep from the nose to the shoulder, giving the appearance of a well-curved and full crest.

The *head* varies considerably in different breeds, in regard to both shape and colour, and, as it is not our intention to describe the features of all breeds,[1] we shall only speak generally. In the Down breeds a fairly broad forehead and a broad muzzle are usually preferred, and, except in the case of the Suffolks, the wool should come well over the forehead and about the upper part of the jaws. A sheep "well woolled" about the head, particularly on the poll, is less liable to injury from flies, which cause great trouble at times, especially if through butting or other causes the skin is broken. The *ears* should not be too thin or papery: on the other hand, they should not be too thick, as this indicates coarseness of skin. The shape of the ears differs with different breeds. There should be no folds of skin under the jaw, as "bottle-throated" sheep are coarse in the skin.

The *teeth* are an important feature, for on the power to graze or gnaw well depends a good deal the amount of food a sheep will get. Short closely set front teeth last longer than long widely set ones ; they are less likely to break, and when old they do not let the grass slip between them so readily.

---

[1] The "standard of points" relating to each may be obtained from the Breeding Association identified with each breed.

No *horns*, or rudimentary horns, are permitted on Down sheep, as they indicate a tendency for the sheep to revert to the unimproved type. Sprigs, or snags, as the rudimentary horns are called, are regarded as serious blemishes. The wool about the head in front of the setting of the ears may contain black hairs, but behind that the wool should be absolutely free from them, as black hairs away from the poll indicate want of selection, and a tendency for the sheep to revert.

Coming back to the body, the *shoulders* should be full but slightly obliquely set; at the top they should be level, so as to give the whole back line a straight and square appearance. If not well buried they suggest the unimproved sheep, and the filling-in behind the shoulders is not complete; this makes a narrow girth round the heart—that is, behind the shoulders. Deepness through the heart is so important that it must never be disregarded.

Both the *fore* and *hind legs* should be set on squarely, neither turned inwards nor outwards too much. A good leg of mutton is an important part of a sheep, as it is one of the most valuable joints on the animal. The legs should be full from all points of view, and the meat should come down well towards the hocks. Squareness of frame when viewed from behind is greatly increased when the legs join well below the tail. If they fork too high up, the legs are not well developed.

The *wool* should grow thickly on the skin, and should be fine in texture and free from dark hairs. Many breeds of Down sheep have been developed from heath breeds which, in an unimproved condition, have a large quantity of black, harsh hairs intermixed with the wool, especially about the hind quarters and the head. These should be bred out, because they are not only injurious to the wool, but indicate insufficient care in breeding and want of thriving properties. When much dark hair is intermixed with the wool on the thighs the sheep are called "breechy."

The *skin* should be clear and healthy; in almost all breeds a clear pinkish skin is best. A dark skin is likely to produce dark wool. Sheep with dark skins are liable to produce snags.

The *underside* of a sheep should be well covered with wool. This adapts them for cold lair. Those with little wool below thrive badly on cold soils, particularly in wet winters. The illustration of the Black-faced Mountain Breed (Fig. 6) shows the more hairy nature of the wool of the Mountain Sheep.

Handling.

"Handling" sheep conveys the idea of touching them to see in what condition they are. The most important points to handle when dealing with ordinary farm sheep with a view to sale or purchase, are the loin, dock, neck, and scrag. The hand should be stretched across the loin, which will show its width and firmness; the fingers should be drawn up to the spine to prove how much it protrudes. The dock should be gripped to ascertain its breadth and fatness; and the "nick," a depression felt for a short space along the spine above the tail, should be found by the fingers, because its size denotes the fatness of the animal, as it is formed by the protrusion of fat on the sides of the spine. If the loin is firm and flat, the dock broad, and the nick well defined, the sheep will "die" well, as these denote that the animal is in a good condition internally, and that the kidney fat is well

developed. The animal may be felt along the spine generally, but except in very highly-fed or show animals, it
is not often that the back is flat right along to the
shoulders, so that it need not be looked for in moderately-
fed animals, although a tendency in that direction is
advantageous; at any rate, the greater the breadth there

Fig. 6.—Mr. J. MacDowall's Black-faced Mountain Sheep.
First Prize, Smithfield, Dec., 1897.

the better, as indicating both the animal's thriving properties and its condition. A grip between the thumb and
the hand will indicate the strength and condition of the
scrag. It is advisable to turn a few sheep to see what
condition they are in on the underside; a brisket well
covered with meat indicates ripeness in the fore quarters;

and if the scrotum is well filled with fat it will prove good internal condition.

Dentition.

A great number of terms are employed to distinguish sheep at various ages. They are so many, in fact, that few farmers are accustomed to all, especially as those in common use in some districts are rarely used in others; therefore, the enumeration of a few is advisable. As some of them are founded on the condition of the sheep's teeth at various times, the dentition should be understood.

For the purposes of the farmer, the *front teeth* or *incisors* are generally sufficient, as it is rare there is occasion to look to the molars or back teeth. But to the exhibitor it is sometimes an important matter to refer to them. There are no incisors on the upper jaw, but in the place of these is a hard elastic pad, as in the case of the ox. The farmer is content to know that the temporary incisors remain in their places until the sheep is rather more than a year old, when the central pair are gradually absorbed by the two permanent teeth.

At fifteen months old the two permanent teeth are well up, and are very distinct from the temporary set, as they are broader and whiter. As at this time shearing generally commences, sheep at this age are sometimes called *shearlings* or *two-teeth*.

At a year and ten months the second pair of front teeth appear well up, having come through the gums at any time after a year and six months. Consequently, if the second shearing is done early in spring, there are two pairs well up, and the animal is called a *four-teeth*.

The next pair come up quicker, and by two years and three months there are three pairs well up; therefore, in late shearing, there are *six teeth*. More quickly the fourth and last pair come up, so that a little before three years are reached the jaw is full, and the sheep is called *full-mouthed.*

The corner teeth do not always come up level until after the expiration of several months. After this, as time goes on, the teeth wear down, and being narrower at the base, they appear to get wider apart.

The *molars* come through at the following periods:—At a month three temporary molars are well up on either side of the upper and lower jaw; at three months the first permanent molar is cut, appearing behind the hind-most temporary molar, and owing to the three temporary molars being afterwards replaced by three permanent molars, is called the *fourth molar.* At nine months the second permanent molar appears behind the first. This is called the *fifth molar.*

At eighteen months the sixth and rearmost permanent molar is cut, and soon after this the first two temporary molars are replaced by two permanent molars, and the third molar is reduced to a shell covering the top of the permanent tooth: before two years the latter will have come through, but will not yet be level with the others. When this tooth becomes level with the others, the sheep is over two years. The dentition of the molars is more regular than that of the incisors, so where absolute accuracy is required they afford the best evidence of the age.

Nomenclature.

The names by which sheep are known vary much in accordance with the locality. In the South of England, where sheep are now brought to the butcher when very

young, the terms used to distinguish older sheep are
becoming more or less obsolete; but in Scotland and in
hill districts generally, where the feeding is not so much
forced, the distinguishing terms are still required as much
as ever.

.When first born, the term *lamb* is universally used, the
mother and lamb being collectively called a *couple*; if twin
lambs, *double couples*. From weaning to first shearing
they are called *hogs*, or *tegs*, tegs being a corruption of
the word tags, as before the wool is clipped the locks
taper to a point or tag; when once the wool is clipped
the locks show a blunt end. In the south-eastern counties there is an exception, the shearlings being sometimes
called *tegs*. The terms *hoggets* and *hoggerels* are used
in some localities.

The males are distinguished by the prefix *he*, and the
females *she*, though the males are sometimes called *wether
tegs*. If the male lamb is not castrated he is called a *ram
lamb* until he is shorn, when he becomes a *shearling* or
two-tooth ram, and afterwards he is called *two-shear*, or *three-
shear* as each year goes by. A ram showing one testicle
is called a *rig*. When the castrated male lambs are first
shorn they are called *wethers*, *shear hogs* (sometimes
, pronounced *sharrig*), or *wether hogs*. In the South
few are kept longer than as shear-hogs, but mountain
sheep are kept on as *three-tooth wethers* and *four-tooth
wethers*.

Female sheep are called *theaves*, *gimmers*, *chilver*, or
two-tooths at the first shearing. After that, as they are
usually used for breeding, the terms *two-shear* and *three-
shear*, or *four-tooth*, *six-tooth*, and *full-mouthed*, are employed in accordance with their age. A ewe which does
not breed is called *guest*, or *eild*; when not in milk, a

ycld ewe; when withdrawn from the flock, a *draft ewe.* The breeding ewes are called *stock ewes.*

Other terms are occasionally used by breeders of and dealers in sheep, but those given in the preceding paragraphs are most often applied.

CHAPTER IV.

THE RAM AND EWE.

The Choice of a Ram—The Ewe—Culling Ewes.

The Choice of a Ram.

THE points which should be observed in buying sheep, already detailed, hold good with respect to the ram. A few other features should be looked for, as the ram is the cheapest means of improving the flock. It is well said that the ram is "half the flock." This does not imply that the ewes should not be carefully selected, as perfection cannot be approached if the ewes are seriously faulty; but fortunately the influence of the ram is very great.

In the first place, a ram influences the offspring of from fifty to one hundred ewes yearly, whereas the ewe only affects her own lambs. It is highly important that the ram should come from a stock which has been carefully bred and selected for many generations. A ram with a good pedigree possesses a fixed type which he imparts to the offspring. The greatest effect of a well-bred ram is apparent when he is mated with poorly-bred ewes: he possesses greater potency over them than he does on ewes which have been well bred for a long time, as the latter have also obtained fixity of type, and he does not make the same impression on them. From this it must not be implied that an inferior ram can be used on high-class

OLD SOUTHDOWN—FOUR-SHEAR RAM.

Bred by Mr. J. Ellman, of Glynde, Sussex. Reproduced from coloured portrait by W. Nicholson, R.S.A.,

ewes without doing harm, as the male is naturally pre-
potent over the ewe; but merely that the extraordinary
prepotency of the well-bred ram over the poorly-bred one
is shown to an exceptional degree. This is well, as it
enables a farmer to improve his flock quickly, and at
small expense. By judicious selection of his ewes, with-
drawing from time to time those which are of inferior
type, the whole flock may be raised to a high standard in
a few years.

In selecting a ram for mating with poorly-bred ewes,
the first point to look for is quality. If the ewes are
small, size must be looked for also, but a squarely framed
ram, with every characteristic of early maturity, must be
sought for. A few pounds extra laid out on a good ram
is money well spent. While, however, recommending that
high-class rams should be used, it is not necessary for
a farmer who intends improving his flock for purely
farm purposes to purchase the most expensive rams, a
few of which are found in most show flocks: these in-
volve an exceptional outlay, as they are required by others
who intend to compete for prizes, and they are a speciality
for which a higher price is paid than a farmer ought to
expend. On the other hand, the inferior sheep of a flock
are not likely to do so much good as those which possess
some slight fault which precludes them from the show
ring but will not be seriously apparent in an ordinary
flock. Cross-bred sheep do not breed so reliably to type
as those which have been kept pure for a long time.

The ram should show vigour, and a masculine character.
This is usually indicated by a strong, though not necessarily coarse, head and a thick neck. The skin is an
important feature, as poorly - bred ewes are, as a rule,
inferior in this respect. The skin should be soft, pinky

and bright. From such a skin wool will generally grow closely and fine. In breeds descended more or less directly from horned breeds, signs of horns should be avoided. A ram should stand well on his legs and move freely.

As a rule a shearling ram is preferred. However, the Hampshire breeders, who have been very skilful in developing their breed, prefer a ram lamb; and they attach much importance to this, as they have found that the use of ram lambs has had a great effect in developing early maturity, for which the breed is so justly famous. A ram lamb should not be mated with more than fifty ewes, and he then will be fit to serve as many as eighty in the next season. If overworked, his successful career will be of short duration; and if a high price is paid for a good ram lamb, at least two seasons' work should be got out of him.

A specially good ram may be used for several seasons —in fact, so long as he is active. After two years, however, there is risk of inbreeding, as his offspring may have been brought into the flock. Where the ram is used for more than two years the ewes should be selected so that mating with his own blood may he avoided. Sometimes, however, inbreeding to a slight extent may be needed to secure fixity of type, or to tone down coarseness, which may have been brought about by injudicious mating or by the soil—some soils having a tendency to make the sheep run coarse.

A shearling ram will take eighty ewes. It is important to purchase a ram in good health, and a ram with foot-rot should be avoided, as it will convey the disease to the flock with which it is put. Should a ram affected with foot-rot be purchased, it should be isolated until all traces of disease have disappeared.

A ram requires to be in good condition at the time of service, but should not be over fat. When got up for sale, rams are often soft, and rapidly lose flesh when turned on to poorer food and put to service; it is, therefore, advisable to get the sheep inured to harder rations by giving them less fat-making foods, but keeping up their vigour with flesh-making foods, and allowing them plenty of exercise. Rams which are strange to each other often fight vigorously at first, and should be watched. Some rams acquire the knack of breaking an opponent's neck. We know of a case where one ram destroyed three others which were consecutively put with him.

When the ram is put to service, his brisket should be rubbed with a mixture of oil and ochre, to show which ewes he has leaped. The colour of the ochre should be changed in a month, so that those which are not in-lamb but come over again may be detected. The time of lambing will then be more accurately ascertained, and as it is often advisable to separate those which are due to lamb early from those which come later, an easy means of distinguishing them is thus obtained. After the rutting season the ram should be taken from the ewes, and may be kept with the wether tegs, where he generally finds a sufficiently good diet. It is not advisable to let him get too poor. He should be kept well on his feet, and his feet should be pared, so as to keep them in good shape.

The Ewe.

The male generally influences the outward form of the offspring, and the female the constitution. It is, therefore, important to breed from ewes of vigorous type. It is not usual to breed from ewes until they are two years old. Often when the stock of sheep in the country is

D

small, lambs are put to the ram so as to produce young at a year old, but this is rarely attended with great success. The difficulty of lambing, and the short supply of milk they are able to produce, together with the dwarfing of size of the ewe itself, are generally sufficient to prevent those who make the experiment from repeating it. Some of the most successful Southdown breeders have been following the practice, and are well satisfied. It is noticeable, however, that those on the South Downs do not do it, and that it is usually done by those on stronger land, where the sheep have a tendency to grow rather big.

The ewe flock, when not directly bought in, should be maintained by selecting the best theaves each year, and putting them into it in the place of those which, from various causes, are drafted from it. It is not customary to run the ewe lambs on expensive lines when it is intended to put them into the flock. A lamb while developing into an ewe has to be kept two years, and if expensively fed throughout that time, is a costly animal. Economy in management must therefore be exercised, but the extreme of insufficient feeding should be avoided. It is advantageous for the ewe to be well grown; consequently, although unnecessary expense is wasteful, she should be kept in a thriving condition. By selecting a draft of the best lambs the flock is improved yearly, especially when the ewes are put to a high-class ram.

An endeavour should be made to obtain similarity of type and features, as sheep, whether sold in a large flock or in small pens of five, always realize a better price when they are well matched in colour and shape of the heads, in size, and in quality of the wool. Quality is the English sheep-farmer's watchword, as it is only by producing meat of better quality than the foreigner that he can reap the advantage which the soil, climate, and better farming of England afford. Let the aim therefore be quality. This is more difficult to get than is size. Too great size, especially in breeds whose special value is in their mutton, is a mistake, as coarse joints are not wanted. Quality is, however, compatible with size, but quality must stand first. Select the ewes with this view. Refuse those which are coarse, gaunt, and narrow in the forequarters. The description of valuable points in a sheep given in Chapter III. should be followed.

Culling Ewes.

After each weaning the flock should be overhauled, so that those no longer worth keeping in it may be taken out to make room for the draft of young ones. The first to go should be the barren or guest ewes; then those which have had diseases of the udder, or are abnormally deficient in their supply of milk. A short supply of milk is not uncommon in ewes with their first lamb, so too much notice should not be taken in the first year. Those which had an inversion of the womb should not be bred from again.

As ewes have to "cut" or graze much of their food, often from bare pastures, or following other sheep on roots, it is necessary that their front teeth should be sound. On stony land the teeth are frequently broken off when they are comparatively new, and they gradually wear away under any circumstances, so that from the time the teeth are fully developed their grazing powers are lessened. When they can no longer graze sufficiently for their proper sustenance, it is no use to keep them in the flock, as the extra labour of producing a lamb brings them to the point of starvation, and either the ewe or the lamb is bound to suffer. The ewes withdrawn from the flock should be sold or fattened off as quickly as possible, according to the food at disposal.

CHAPTER V.

THE GREEN FOOD SUPPLY.

Suitable Crops.

THE simplest system of sheep-keeping is that by which the animals are fed on grass throughout the year. Here the supply of food is dependent on the amount of grass grown, and varies with the productiveness of the season. The system most requiring skill and forethought is where the sheep are kept entirely on crops of arable land, while the mixture of grass and arable land crops is midway between the two.

The crops which are most serviceable on arable land may be divided according as they are suitable for particular times of the year. It is usual to divide the sheep year into two portions, one commencing at the beginning of November and lasting until the end of April, known as the winter season, and the other occupying the remaining portion of the year, known as the summer season. The winter is the season of root crops, the summer of grass, clover, and other green crops. The same kinds of roots are not so valuable at all periods of the winter season, and a succession of crops suitable must be arranged.

By dividing the year into the following periods, the crops which are most reliable, and are best suited in ordinary seasons are readily seen, though, of course, they overlap to some small extent.

37

November to February.	March to May.	May to August.	September and October.
Swede turnips	Mangols	Clover and seeds	White turnips
White turnips	Kale	Grass	Drumhead cab-
Drumhead cab-	Winter greens	Trifolium	bages
bages	Early catch	Early cabbages	Late rape
Kohl rabi	crops, such	Early rape	Stubbles
Grass	as : Tares	Winter and	Grass
Ensilage	or vetches,	summer	Young seeds
Late-sown tur-	winter rye,	vetches	Hay
nips run to top	winter bar-	Sainfoin	
Hay	ley, winter	Lucerne	
	oats		
	Water mea-		
	dows		
	Late-sown tur-		
	nips run to top		
	Rape		
	Hay		

Swede turnips should not be fed before November, as, although they may have acquired size, they are not ripe. Unripeness causes scour, and many sheep are lost by being put on to them too early. The food contained in them is also not in a form which can be fully utilised. From November to March is the season of the swede. After March they become dry and pithy, much of their feeding value having been lost.

Kohl rabi cover much the same period as the swede, the Early Small Top variety being excellent early winter feed ; the Late Big Top, being hardy, is valuable at a later period, as it withstands frost. When planted, care should be exercised to obtain the variety suitable for the season at which the crop will be required.

Mangels are essentially spring food, although they are frequently fed at an earlier period. It is wasteful to feed them early, as the food is not wholly digestible. The

greatest value is found in them from March to June, if
they can be kept so long; the greater part of the food in
them is then made use of by the animals. When used
for male sheep in the spring, the yellow shoots should be
stripped off, as they tend to produce urinary troubles by
forming crystals in the passage from the bladder, which
check the flow of urine, and cause inflammation of the
bladder, often resulting in death. Even without the leaves,
there appears to be a tendency for these crystals to form,
but to a far less degree. Ewes, being possessed of a
bigger passage, rarely suffer.

Cabbages supply a succession of food from June to
Christmas, if early varieties are transplanted in autumn,
and later varieties of Oxheart type are transplanted in
spring. These will be available until September, when
the Drumheads, also transplanted in spring, or drilled and
set out very early, will be available until the end of the
year; but they cannot be relied upon longer, as when
thoroughly ripened they are susceptible to destruction by
frost.

Kale, such as Thousand-head, Shepherd's, and other
hardy varieties, have a special value as coming to their
best between the early part of March and May, supplying
rich, succulent food during the hard pinch between roots
and grass. Shepherd's, when planted early, produces a
heavy crop early in autumn and winter, owing to its
exceptionally thick leaves and rapid growth, but does still
better when left until spring for a greater quantity of
side shoots to push out. The tardiness with which
Thousand-headed kale comes on at first prevents many
from using it, but it crops so well when it has been on
the land a year that it is worthy of a more extended
cultivation. The advantage of both these kales is that if

a cutting is taken in April, another crop will be available in July.

Rape is of two types, Giant and Dwarf. The Giant produces a single crop, and if sown in May or June is fit to feed in August and September, when it attains great size. The Dwarf kind may be sown from March to July, and will produce several crops before it is necessary to destroy it. March-sown rape will often produce good feed in mid-June, another in September, and valuable green food in spring. Lambs should not be put on it a second time unless it has been well frosted, as it becomes "sour," and then causes scour. As a rule it is sown at the same time as white turnips, and then gives an autumn and a spring crop. It is valuable as a catch crop taken off land from which an early summer crop has been taken. It then makes excellent lamb food in spring.

White turnips are available from August to December. Those sown as a catch crop in July and August are useful in spring, when the tops provide good lamb food. Quickly-maturing varieties, like Stratton Green and Six-weeks, are valuable in seasons when the main crop of roots has failed. Sown in July and early August, they make useful autumn feed.

Winter greens, such as Savoys and Brussels Sprouts, are very hardy, and stand the severest weather in winter.

Rye, winter barley, winter oats, winter vetches, and *trifolium* are essentially autumn-sown catch crops, available at the critical period after the root crops are consumed. Rye is the first to be fit for feeding; vetches and trifolium come a little later. A small piece of *winter barley* is valuable on all farms where lambs are raised, as it is one of the best antidotes to scour, the great scourge of the lamb-breeder.

Ensilage is of great value to the sheep-farmer, and silage is especially valuable as food for ewes, as they eat it readily, thrive well, and it produces a good flow of milk. It is a convenient and most useful fodder in long-continued frosts, when roots get frozen into the ground, and are difficult to obtain in sufficient quantity. The comparative warmth of silage is also a matter of great consideration. Frozen roots are indigestible, and sheep thrive badly on them, as the goodness they contain is expended in maintaining animal heat within the body, and little goes towards improving the condition of the sheep. We have a high opinion of the value of silage, and strongly recommend its more general use. An expensive silo is not necessary, a field heap being economical in all respects except outside waste. Clover and other seeds make good silage, and if made in the field where grown, the cost of carting to the silo is small.

Sainfoin, clover, lucerne, grass seeds, and similar crops grown as short leys are the mainstay of the sheep keeper on arable land during summer, and in one way or another they may be relied upon from May to November. They are also the chief source of hay for winter feeding.

Stubbles of corn crops afford sweet fresh food, and are specially valuable when they contain strong, young seeds.

Pasture affords keep at almost all seasons.

Water Meadows have a special value as affording a bite of fresh grass as early as the first of April.

Catch Crops are of so much importance to the sheep farmer that more than a mere notice of their value should be given. They may be regarded as being specially valuable on two occasions—in time of drought, to augment the limited amount of sheep food during the coming months; and in ordinary seasons, for utilising the land

when it would otherwise be lying idle and losing its
manurial constituents. Crops which are suitable for sowing
to produce keep within the shortest time are mustard and
stubble turnips. In dry seasons the land earliest cleared of
corn or other summer crop should be brought into a good
tilth, and be sown with one of these to produce early
autumn keep. It should never be forgotten that every
day is of value, and that no time should be wasted in
getting the land sown, whatever catch crop is taken.
Hardier varieties of turnip, kale, and rape, sown as catch
crops, will produce excellent feed in spring, and be
particularly valuable for lambs. The autumn-sown catch
crops, suitable for feeding in late spring, such · as large
winter barley, oats, vetches, and *Trifolium incarnatum,*
should be sown as early as possible. The cultivation for all
but the trifolium consists of nothing more than ploughing,
sowing, and harrowing, a very simple seed-bed being
sufficient. Trifolium thrives better when the land is not
ploughed, but merely harrowed.

Hay is referred to in the next chapter.

CHAPTER VI.

CONCENTRATED FEEDING STUFFS.

Selection of Suitable Foods.

THE most valuable feeding stuffs are linseed cake, cotton cake, special feeding cakes and mixtures sold for sheep food, wheat, barley, oats, peas, beans, maize, lentils, pea husk, malt, malt culms, bran, rice, and linseed. Fenugreek, ginger, and other spices are employed to give an aromatic and enticing flavour to other foods. The terms *cake* and *corn* are used somewhat indiscriminately in some districts, farmers frequently saying that they are giving their sheep so much "cake," when a portion of it is "corn," or so much "corn" when some of it is "cake." What they mean to imply is that they are giving a certain quantity of rich concentrated food.

Linseed cake is the best individual concentrated food. It contains flesh-forming and fat-forming constituents in good proportions, and the oil acts beneficially on the bowels. The ease and safety with which it can be given makes it very popular with flockmasters, and this tends to keep the price somewhat higher than that of a mixture of other foods of the same value, which can be prepared by any one knowing the constituents of cake, and of the other articles on the market.

Cotton cake is sold in two forms, known as "decorticated " and " undecorticated." The former is freed of husk,

43

and in the latter the husk is left. The feeding value of the decorticated is much greater than that of the undecorticated; it is richer in feeding constituents, and is not so astringent. It should be broken very small, as almost all samples are hard; those containing hard, brown patches varying in size from a bean to a half-crown in circumference are particularly dangerous, as these pieces are almost indigestible, causing great irritation in the stomachs of old sheep, and frequently prove fatal to young sheep. These hard pieces must therefore be broken, and for this reason the cake should not be given in lumps larger than a bean. All cakes should be broken small, but it is of the greatest importance in the case of decorticated cotton cake. As cotton cake contains a large proportion of flesh-formers, it should be given in mixture with foods containing more starch.

Special feeding cakes are usually well compounded and safe foods. This is not invariably the case, and sometimes very inferior stuff is employed in their manufacture. Such foods should be bought on analysis, and care should be taken on delivery that they are sweet and in good condition. Mouldy cakes of any kind are poisonous.

Wheat, barley, and oats are valuable foods, and at the present low prices are economically used as sheep foods. As a rule they are better for being mixed with peas, cotton cake, or other nitrogenous foods. With all grain and starchy foods a small quantity of crushed linseed is of special value, as the oil greatly aids digestion and helps to maintain the health of the animals. By themselves they are found to be heating.

Peas, beans, and lentils are excellent sheep foods, helping to form a good proportion of lean meat, and making the sheep handle firmly. It is important that peas and beans be not used when they are new, as they are then indigestible, causing scour and other disturbances, and do not yield their full feeding powers. They are not considered old until the March following their harvesting, and it is better to let them remain a full year before being consumed. Peas, beans, and all corn and grain should be passed through a kibbling machine. It is advisable not to reduce them to a fine meal, but to crack or grittle them, as meal is liable to be blown out of the feeding troughs, or in wet weather to form a paste. Large quantities of unbroken corn swell sufficiently to distend the stomach unduly; and the full value is rarely obtained when given whole. The cost of grittling is well repaid.

Maize and rice are starchy foods, well adapted for mixing with the more nitrogenous foods.

Malt culms, pea husk, and bran are valuable foods, and have a special utility for mixing with chaff or chop to make animals eat a larger quantity. Pea husk is very suitable food for lambs, and we know of no other food which keeps them in such good health.

Hay, if not a highly concentrated food, is a valuable aid to the sheep keeper, as, in addition to a fair amount of feeding properties, it tends in no small degree to keep the animals in a thriving and healthy condition. Sainfoin, lucerne, broad clover, and "mixture" from temporary leys are, if anything, preferable to meadow hay, although the latter is very valuable. Sainfoin and broad clover are the best. It may be given in racks in ordinary condition, or in troughs chaffed.

CHAPTER VII.

SECTIONS OF THE FLOCK.

Composition of the Flock—General Arrangement of Sheep on the Farm—Head of Sheep at Different Seasons—The Shepherd's Account Book—Sheep required to Feed off the Root Crops.

Composition of the Flock.

THE previous chapters have been devoted to matters which have an important bearing on sheep management, and they should be sufficient to enable the reader to follow with ease descriptions of systems of management which will be given in the subsequent chapters. The most important sections of sheep farming are included under the following headings :—

Management of Ewes.
Management of Lambs.
Management of Stores and Wethers.
Management of Fattening Sheep.

These are subject to many subdivisions, and an endeavour will be made to lay before inexperienced sheep-keepers. the work of the flockmaster from day to day and from season to season, according to the systems of farming which may be practised. As grass sheep-farming is not so complicated a calling as arable sheep-farming, attention will be mostly directed to the latter, especially as the proper

LINCOLN SHEARLING RAM.

management under this heading includes most of that under grass-farming.

Under ordinary circumstances a flock is composed of ewes, lambs, store and fattening tegs, and wethers. The ewes are regarded as the scavenging portion of the flock, as, except at lambing time, they clear up behind other sheep, and are run as inexpensively as possible consistent with their healthy maintenance. Lambs which are to be fattened off as fat lambs, or tegs, are kept on the best and freshest food, and generally receive cake or corn in addition. Ewe lambs to go into the flock are not forced, but are kept on sweet fresh food. Where an endeavour is not made to get out sheep until they become wethers, they are not fed highly or expensively kept until some little time before they are fattened out. A chapter will be devoted to the management of a show breeding flock, but it will be better not to deal with them here, as they form a distinct feature in sheep-keeping.

General Arrangement of Sheep on the Farm.

Perhaps the best way of showing how a flock is managed is to take an imaginary farm of mixed grass and arable, and show how a flock of 200 ewes and their offspring may be maintained on it. The size of the farm need not be definitely fixed, as on such a farm cattle are always kept, but sufficient grass and arable must be allotted from time to time to maintain the sheep at various seasons. Different soils possess varying powers of carrying sheep, as may be gathered from the fact that on some of the arable sheep farms, with a small proportion of grass and water meadow, and with a mixed cropping of corn and fodder crops in which the fodder crops embrace the larger acreage, from two to two-and-a-half sheep per acre over the whole acreage

of the farm are sometimes permanently carried (this is
particularly the case where an extensive system of catch
cropping is adopted) ; whereas on the thinnest hill land
and heath some acres have to be given up to one sheep.
On a mixed farm in the Midland counties, with a small
proportion of grass, but with the arable land fairly
divided among corn and fodder crops, one to one-and-a-
quarter sheep over the whole farm may be carried, though
this number is found oftenest in winter, when they are
on the roots, an extra number of tegs having been bought
in for wintering.

The best place to take up the treatment of a mixed
flock is where the ewes have been deprived of their lambs,
as then the whole of the breeding and management of the
lamb until he is sold lies in the future. The stock is
then complete also.

The season of *weaning* varies in accordance with the
period at which the lambs are born, and is generally
about four months from the time of birth. Where lambs
are born very early, however (irrespective of show animals),
it is customary to dispose of the earliest as Easter fat
lamb, the sales continuing for some time. Dorset lambs,
which are bred in the autumn as early as October, are
sold in early winter. Dorset ewes (see Plate) are put to
the ram in May; and often breed twice in one year, being
put to the ram a second time as soon as they come into
season again. Hampshires and Oxfords are born in
January, February, and March ; Shropshires and Southdowns in February and March ; and white-faced sheep
in February, March and April, according as the grass is
likely to come in. The *lambing season* is practically con-
trolled by the season when suitable food can be obtained,
and the time when grass will be ready for the lambs is

the chief regulator. On arable land, however, the cropping
can be arranged so as to be available at any time, and
the time of grass need not be regarded so closely. On
hill land and mountains the lambing time is sometimes
deferred as late as June. For the purposes required . here,
as affecting sheep on a mixed farm, February and March
may be taken as the lambing season; consequently, the
weaning season may be put at June.

Head of Sheep at Different Seasons.

A certain loss is always experienced, and it usually
varies from three to five per cent., but from abnormal
circumstances may go far beyond this. However, in the
estimate given here they are not taken into consideration,
round numbers being adhered to.

Taking the flock of ewes as being 200 at the time of
weaning, they may be estimated as having brought up
250 lambs, fifty of which have been sold as fat lamb.
There should be fifty theaves ready to come into the flock.
Lambs weaned at this time in the previous year should
not be kept on to be fattened as wethers : therefore, in
this case they are considered as having been sold, though
there are instances where they might be run on to be
fattened on the seeds during summer. The stock is,
therefore, 200 ewes, 200 lambs, 50 theaves ; a total of 450.

Immediately after this fifty ewes are culled from the
breeding flock, and may be sold at once, or be fattened
during summer. The chance of selling the ewes affords
an opportunity of lowering the number to be kept through
the summer, should the prospect of keep be poor. The
food available consists of seeds or leys, vetches, and early
cabbages. On a mixed farm it is usually advisable not to
stock the pastures until after they have been mown.

E

The same number of sheep, less the. fifty culls, should
be kept through the autumn, and it depends on how well
the tegs have thriven when they are fit to be sent to the
butcher. As fat tegs, the earliest rarely go to market
before the end of November, so the winter season is entered
with the same number. During autumn the ewes will
have found food on. the stubbles and aftermaths of grass
and leys, together with a little stale food behind the leys
or cabbages and white turnips. The lambs at the same
time may have had the first pick over the stubbles, grass,
and leys, and the first gnaw at cabbages or white turnips,
with a small quantity of cake or corn.

The winter season (from November to May) sees great
alterations; the tegs are gradually sold, and a new crop
of lambs appear. Half the tegs may go before the lambs
are born, and the other half, minus fifty she tegs which
are to be kept on as breeders, may be sold subsequently.
This allows for thirty to go in each month from December
to April inclusive. It will, however, depend very much
on the quantity of cake they have received from weaning
as to when they are fit to go out, and if it has been
moderate, the whole 150 may go out between February
and April inclusive. This has to be borne in mind when
arranging the cropping.

More briefly put, there would be at the weaning
season, 1898, 200 ewes, 200 lambs, and 50 theaves on the
farm, making 450. However, 50 cull ewes would soon be
sold, leaving 400. At the 1st of January, 1899, the same
sheep remain, less 30 fat tegs sold, leaving 370.

During January, February, and March the remaining
tegs, less fifty ewe tegs to be retained for the flock, are
sold, reducing the number of sheep to 250. Meanwhile
250 lambs are born, so that the total is 500; but fifty of

these are sold fat in spring, bringing the total at weaning time, 1899, to 450, as at the same period in 1898. This is regardless of sheep which may have died or have been purchased.

The Shepherd's Account Book.

It is important to keep a strict record of the sheep, as with fluctuating numbers there is no other way of proving loss. The shepherd should keep a note-book, in which he should record all sales, losses, and purchases; he should also note the number taken into each field, and the number removed from the field after the food is consumed, so as to make a check by which he can prove the loss by straying or stealing of any sheep in his charge.

The farmer should check his book, and post it in his sheep account. He will find on page 52 a useful form in which to record the sales and purchases of sheep, also the births and deaths.

Sheep Required to Feed off the Root Crops.

As a rough estimate of the carrying powers of an acre of cropping, it may be taken that a teg having an allowance of corn, eats its own weight of swede turnips each week, and an extra sixth more of white turnips. An ewe will eat at least half as much more. Thus, taking the average of the tegs to be eight stones, each sheep will eat 1 cwt. per week; and each ewe 1½ cwt. A twenty ton crop of roots, which gives 400 cwt., will therefore keep 200 tegs for a fortnight, and 200 ewes three-quarters of a fortnight.

In arranging for the feed until May, the eating capacities of the earlier lambs have to be taken into consideration. The earliness with which they are born of

THE SHEPHERD'S ACCOUNT BOOK.

Sheep Purchased.

Date.	From whom.	Class of sheep.	Number.	Price.	When received.	When paid.	£ s. d.
1890.							
Jan. 1	In stock	Ewes	198	55s.	—	Valuation	544 10 0
Jan. 1	In stock	Tegs	146	44s.	—	Valuation	321 4 0
Jan. 1	In stock	Rams	8	90s.	—	Valuation	12 0 0
Feb. 12	Bred	Lambs	140	—	—	—	—
Mar. 1	Bred	Lambs	187	—	—	—	—
July 28	R. Jones	Wethers	50	37s.	July 28	July 28	92 10 0
Aug. 7	R. Smith	Ram	1	10 guineas.	Aug. 17	Aug. 17	10 10 0

Sheep Sold.

Date.	To whom.	Class of sheep.	Number.	Price.	When delivered.	When paid.	£ s. d.
1890.							
Jan. 14	By auction	Tegs	20	45s., less 10s.	Jan. 14	Jan. 18	44 10 0
Jan. 21	W. Hill	Tegs	25	50s.	Jan. 24	Jan. 28	62 10 0
Jan. 23	Died (S. Brown)	Ewes	3	5s.	Jan. 10, 21, 28	Feb. 17	0 15 0
Jan. 27	Died (S. Brown)	Ewe	1	7s.	Jan. 28	Feb. 17	0 7 0
Feb. 5	W. Hill	Tegs	15	47s.	Feb. 6	Feb. 8	35 5 0

course greatly controls the quantity of food they will eat; and this food will rather be an equivalent to roots, than roots themselves, as they require soft succulent food, such as turnip tops, rape, or kale. All food, however, is best estimated on the equivalent of swedes, or other standard crops. Mangels, kohl rabi, and cabbages (as long as they last into winter), may be regarded as the equivalent of swedes. Sprouting crops are more difficult to estimate, as the severity of the winter and the backwardness of spring growth affect them. The feeding power of a kale crop may far exceed that of roots even in winter time, and to a greater extent in spring; but this only relates to that planted early in the previous spring. Late planted kale and rape give small return in winter; but when kale gets its spring shoots in April, it crops very heavily.

It must always be borne in mind that there is greater likelihood of decrease than increase in the supply of food, as the risk of injury from frost, mildew, and other destroying powers has to be run; therefore, it is unsafe not to make the provision per head in excess of bare estimates. The strain on the food will be heavy in the early spring months, when the 200 ewes, and a portion (say half) of the tegs are still in hand, and the lambs are beginning to feed. All sections of the flock require to be well fed at this season: the ewes to supply milk for the young lambs, so that they may not receive a check in growth, the togs so that they may come rapidly into condition for killing.

The food available during March and April consists of a portion of the swede crop not consumed, mangels, autumn-sown catch crops (except tares), rape and kale. In May the leys and grasses may be looked to for help; trifolium and rape, together with autumn-sown catch crops, including

tares, should also be available. The fat tegs should have
been sold; but the place of the latter portion sold is more
than taken by the theaves and young lambs, which by
this time have acquired big appetites. Where the land
is largely composed of pasture, or where grasses and clovers
are grown in rotation, the flockmaster can generally feel
secure of feed throughout May. Details of management of
each section of the crop are given in subsequent chapters.

If the food of the farm shows at any time that there
will be more than is required to carry such a flock as
has been described, the farmer has a chance of buying in
more, if the prices look favourable. It is, however, foolish
to buy sheep which show no chance of profit. It is better
to take in others at agistment, at a small price per week.
Failing this, it is wiser to plough in a portion of the crop
as a green manuring. By doing this, the land gets the
advantage of earlier working, which is always beneficial.

OLD LINCOLN RAM. *Reproduced from coloured portrait by W. Nicholson, R.S.A., published in 1841.*

Bred by Mr. Jex St. Jermains, of Lynne.

CHAPTER VIII.

A YEAR WITH THE EWES.

The Ewe at Weaning Time—Treatment of Ewes before and during Gestation—The Lambing Yard—The Shepherd's Requirements in the Lambing Pen—Lambing, or Yeaning —Malpresentations—Prevention of Infectious and Contagious Diseases in the Lambing Pen—Daily Work in the Lambing Fold.

HAVING given a brief description of the position of each section of the flock at various seasons of the year, the details of management can now be gone into. The ewes are the first section to deal with, as they are the most permanent portion of the flock, and as the lambs and future sheep are dependent on them, it is most convenient to treat of them.

The Ewe at Weaning Time.

Taking up the subject of the ewe at the time the lamb is weaned, the first thing to attend to is to see that she does not suffer from weaning. Attention should be given to the udder, to see that it is not unduly inflamed or distended. A point is always gained when the ewes are shorn before weaning, as then a casual observation of the udder reveals its condition. When the wool is overhanging, it is necessary to catch the sheep and handle the udder. If milk continues to form in large quantities, the udder should be drawn a little to ease the distension, but it is not necessary to empty it. A dose of Epsom salts will help to relieve it, and may be used without fear of injury. Succulent food should be withheld, and the diet should be somewhat scanty and poor. In case inflammation sets up, the same treatment must be more rigorously adhered to, and the udder should be well rubbed with lard. If care is taken at the outset, serious complications rarely arise.

No time should be lost in getting the flock overhauled; those not suitable for breeding from again should be culled in accordance with the instructions given in Chapter IV. The culls, if not sold at once, should be put on to good feed, and, if necessary, be given cake to fatten them, as no good end is gained by keeping them on the farm longer than there is absolute need of. They may be sold off as they become fit for the butcher, or all at once, as appears best. No other special treatment is necessary.

In some of the large breeding districts the ewes are sold off as soon as they have produced and reared a lamb after becoming full-mouthed, and are then bought up for breeding in other districts, where the custom prevails of buying in such ewes with a view to letting them breed once more, and then fattening both the ewes and the lambs simultaneously—a practice which is often attended with considerable profit.

The ewes having been drawn out, should be branded with a pitch brand. For the better identification of the sheep, it is advisable to mark all of them in special parts, according to their age and sex. Various systems are adopted, but nothing is simpler to remember than the following, which embraces the ordinary sections of the flock :—

Ram tegs, on the off shoulder.

Ewe tegs, on the near shoulder.

Theaves, in the middle of back on near side.

Wethers, in the middle of back on off side.

Two-shear ewes, on hip on near side.

Three-shear ewes, above tail.

Four-shear ewes, on hip on off side.

A special effort should be made to get the feet sound, as effects of the yarding are often in evidence where proper care is not taken. This is one of the most leisurely periods of the shepherd's year, and advantage should therefore be taken of it in this way; also in mending hurdles, or getting forward with anything which has been allowed to drop in arrear during the previous busy season.

Treatment of Ewes before and during Gestation.

The breeding flock should be got into good condition to receive the ram, as then a greater fall of lambs may be expected in the following spring. Ewes falling in condition do not come into season so early, and are less likely to produce twins. If the ewes do not come into season at the desired time, a little heating corn, especially barley, will help to bring them on. Barley and wheat stubbles are useful in this respect. After being put to the ram, the ewes may be kept moderately well; it is best not to allow them to become fat. Grass is a very economical food during the autumn and early winter, but they may be put on the arable land at night to clear up food which other sections of the flock have left behind.

When the wool gets long enough to carry the dip, ewes should be dipped in one of the recognised dips, otherwise they will be seriously troubled by vermin, and in spring the lambs will catch the pests from them, much to their detriment.

Ewes go with young about twenty-one weeks, but this varies slightly, as does the period of gestation in most animals. During gestation the ewes should · be allowed plenty of exercise, and this advantage is gained when they roam on broad pastures. Grass land is eminently suited to ewes in lamb, but when food gets short, especially during the last few weeks before lambing, they should receive extra food.

The idea commonly prevails that turnips exert a baneful influence on the ewe and the lamb she carries. That ewes fed exclusively on turnips do give trouble at lambing, and often cast their lambs prematurely, there can be no doubt, but it is less from the presence of injurious matter contained in the roots than from the want of nutritive matter. The ewe for a few weeks before lambing has to build up the lamb within her, and this is a great strain on her. If she is in low condition and is supplied with insufficient nutritive food, Nature revolts, and either the ewe or the lamb suffers. The lamb most often suffers, and is expelled before time, Nature taking this course to save the ewe. The same result is obtained if ewes are kept too long on poor grass. We had experience of this some years ago, when, owing to a shortness of keep, we sent our ewes to graze at a distance. They stayed too long, and our losses at lambing were the heaviest we ever experienced. Our district being a strictly arable one, ewes rarely go on grass, yet by giving a plentiful supply of dry nutritive food with the roots, they are as prolific as in any other district.

The fact is, turnips contain little more than sugar and water, and an ewe cannot support herself and build up a lamb on such a thin syrup; it is not natural. She requires a more complex food, containing that which will supply the means of building up bone, muscle, and other parts of the lamb, without unduly exhausting herself. This must be borne in mind, and as the greatest strain comes when parturition approaches, care must be taken to give the ewe good food. At the same time the ewe should not be fat. She should be vigorous and muscularly strong. Dry food, such as hay, pea-haulm, or straw-chaff, with malt culms or other nitrogenous food, is well suited for this. A small quantity of peas, decorticated cotton cake or similar concentrated food may be given with much advantage, if coarser food cannot be conveniently spared. Corn of a starchy nature, such as maize, is not so valuable. It is a mistake to delay giving the ewes dry, bulky food until close upon lambing time. Hay or other dry food should be given in early autumn, when many cases of premature lambing would be prevented.

Care must be taken that as lambing time approaches ewes are not chased by dogs, or in other ways made to hurry themselves unduly. Ewes which are close-folded should be taken out for gentle exercise for a mile or two daily, special care being taken that they are not hurried through gateways, as the concussion is injurious to their young. Ewes—in fact, all sheep—should be approached gently, so as not to frighten them. Strange dogs should be kept well to heel.

When ewes are well gone in lamb, they must not, if foot-rot is present, be turned to dress their feet; but the feet must be lifted carefully and dressed or pared if they get badly grown.

On those farms where permanent lambing folds are not found, it is necessary to provide proper shelter for the

ewes and lambs during the lambing season. This should
be in readiness before the most forward ewes are expected
to lamb. Where temporary lambing pens have to be put
up, they should be placed in the most convenient position
with regard to food, so as to avoid unnecessary carting.
This should be decided upon long before, and to provide
straw for litter a stack of corn should be built near, so
that when threshed the straw may be available. There
is no better form of temporary lambing pen than that
adopted on the Wiltshire and Hampshire Downs. Except
on small or very compact farms such yards are preferable
to a permanent one, as the carting of bulky material to
and fro, and the expense of getting the manure made in
the yards on to the land subsequently, is avoided, and
well repay the trouble of erection.

The Lambing Yard.

Before describing the method of constructing a yard, it
is advisable to point out what divisions of an ordinary
lambing flock are usually found necessary. In dealing
with a ram-breeding flock, many subdivisions of the lambs
according to sex and age are necessary. The ewes require
to be handled to see which are likely to lamb earliest,
which is generally indicated by the udder, though young
ewes frequently do not show much sign of milk until just
previously to lambing. Those likely to lamb at once should
be separated from those which will not lamb for some
time; drafts of the latter should be put with the former
from time to time. Those which will lamb early should
be yarded at night, but it is advisable to keep the others
out of the yard as long as possible, because foot-root is
likely to be contracted in the yards, as the feet soften
and become foul when kept on dirty bedding.

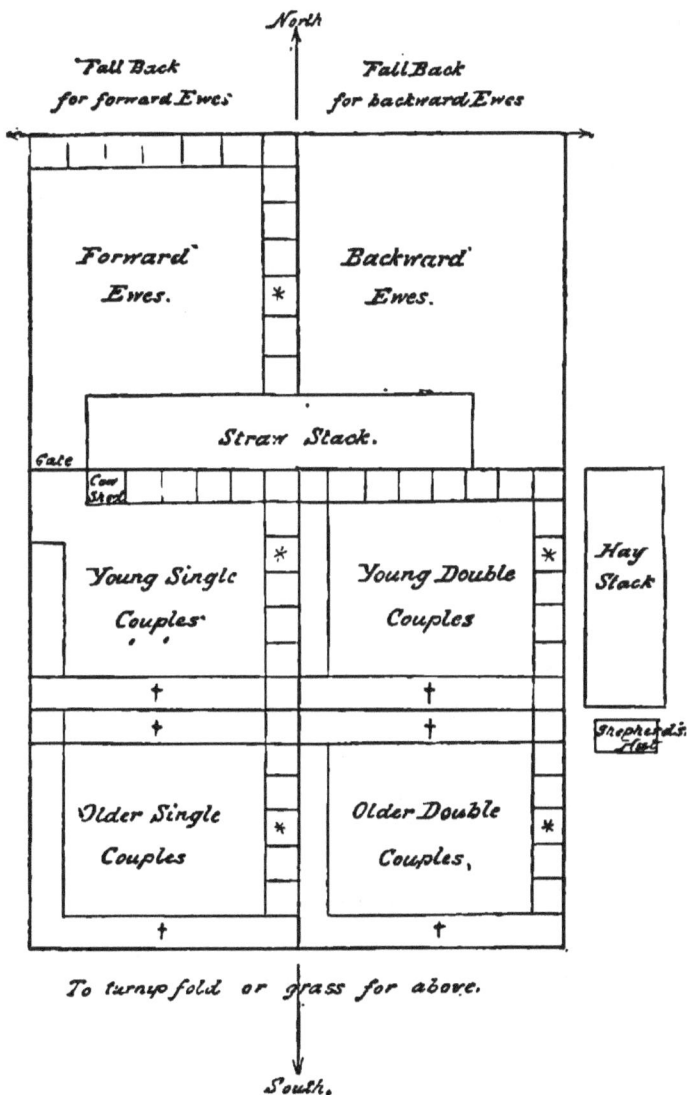

North

Tall Back
for forward Ewes

Tall Back
for backward Ewes

Forward
Ewes.

*

Backward
Ewes.

Straw Stack.

Gate

Cow
Shed

Young Single
Couples

*

Young Double
Couples

*

Hay
Stack

†

†

†

†

Shepherds
Hut

Older Single
Couples

*

Older Double
Couples,

*

†

†

To turnip fold or grass for above.

South.

Fig. 7.—Ground Plan of Field Lambing Fold.

* Ewe and Lamb Pens. † Shelter along Sides of Fold.

Around the ewe yard, a number of small pens, a hurdle square, should be fixed, so that, as lambing becomes imminent, the ewes may be separated. A pen should also be provided for the ewes which have lambed, and in large flocks additional yards are required, so that, as the lambs become stronger, they may be taken out into the fields during the day. The divisions are, therefore, made, in accordance with the size of the flock in the first instance, and the age of the lambs subsequently.

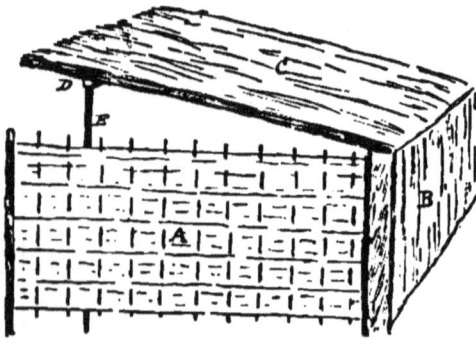

Fig. 8.—Section of Outside Shelter in Lambing Pen.
A, Dividing Hurdle; *B*, Back Rows stuffed with Straw; *C*, Roof; *D*, Supporting Batten carried on Fir Post (*E*).

It is a good plan to build the yard so that the straw stack is in the middle, to facilitate rebedding (Fig. 7). The method of making a yard as carried out in Wiltshire and the neighbouring counties is as follows: Rows of hurdles are set up to form the outsides of the pens; about 4ft. inside these, posts about 6ft. or 7ft. in length are driven into the ground at intervals of about 10ft. Deal battens about 3in. by 2in. are nailed from post to post to afford support to the roof. Outside the row of hurdles, but close to it, another row is laid on the ground, and on this a layer of straw is placed ; these are then lifted up, so as to stand parallel with the first row, being held in position by stakes. In doing this, the straw becomes held up between the double rows, and effectually blocks out the wind. Hurdles are then laid from the top of the back row to the line of battens to form a roof. A covering of straw is placed on these, and on this another layer of hurdles is laid and secured. A substantial roof is thus obtained in a very short time (Fig. 8 illustrates a section).

This shelter should be placed round all the pens, and a division to contain each ewe and lamb for a few hours after lambing should be provided by making partitions at each hurdle's length along the north and east sides. A line of hurdles placed at right angles to the row will do this, and the front of the division can be closed by another hurdle. Each division will then be a hurdle square. A few only of these divisions need be provided for the older lambs, where they are useful in times of sickness. When the lambing season advances and the lambs require less shelter, the straw from the walls and roof is utilised, the destruction of the yard being bought about gradually.

In other districts other methods of protecting the lambs are adopted but the one next most serviceable to that described is that used on the large farms in the eastern counties, where, instead of stuffing the hurdles in the way described, two rows are set up about 3ft. apart, parallel to each other, and at right angles to these other hurdles are set to form small pens. The space between the double row is filled in with straw, and to afford shelter, poles are laid on the dividing hurdles reaching half-way or more along them. The straw is spread over

the poles, and formed into a roof which is roughly
thatched. The principle of the protection is thus a long
narrow stack, with the eaves carried over to a considerable
degree. The straw used for protection becomes available
for litter subsequently. Rows of bushes or wood are
occasionally set up to break the force of the wind. The
use of sheltering cloths attachable to hurdles (Fig. 9) has
come into vogue of late, and they form an efficient and

Fig. 9.—Sheltering Cloths.

easily erected shelter, whether in the lambing pen or in
the open field.

In arranging the yard (Fig. 7), the ewes which have not
lambed may be kept on the north side of the straw stack,
and the young lambs placed on the south side. The straw
stack is best built long and narrow. A hay stack should
have been built near, and the shepherd requires a portable
hut in which to sleep and to keep medicine and corn.

When the ewes are short of milk, a cow kept in the yard is useful.

Ewes require a good "fall back" or space on which to rove. It is not always convenient to have this attached to the lambing pen, but it should be arranged that where they feed, even if a fresh piece is given daily, the hurdles are not kept within too close quarters. It is important to place the fold on a dry, solid field, an old ley being very suitable; if on a hill-side facing the south, so much the better. Any amount of subdivisions may be added to those shown in Fig. 6, if found advisable. A fair number of subdivisions make the shepherding easier, as the sheep are more quickly handled.

The Shepherd's Requirements in the Lambing Pen.

A soldier cannot take the field unarmed, and a shepherd equally requires appliances, lotions, etc., to help him battle with the difficulties before him. The cow will supply the milk necessary for the lambs, which otherwise would not get sufficient. The cow should be neither too freshly calved nor too stale. The hut should be provided with a small stove to supply warmth, care being taken that the fumes can escape readily, as many shepherds have been suffocated by closing ventilating apertures on cold nights. The stove will heat the milk, warm the necessary water, fry the shepherd's rasher, and warm his billy, whatever it may contain. The vessel in which the milk is boiled should be made on the glue-pot system—a smaller one inside a larger one containing water, to keep the milk from burning.

Among other things which he will have to provide are a ball each of string and stout tape. In those cases where it is desired that the string shall not slip, tarred cords

F

are convenient. These are required for securing to the legs of lambs during malpresentations, for binding the side of the shape in cases of eversion, and for many other purposes during the course of the lambing season. Ruddle or ochre of several colours is required to make distinguishing marks. In pedigree flocks ear-markers may be used with advantage, to prevent ultimate confusion.

The shepherd will also need one or more drenching horns or bottles (an old sauce bottle with strong, long neck is very convenient); a cordial for chilled or weakly lambs (equal parts of brandy and sweet spirits of nitre); a bottle of diarrhœa or scour mixture (Mr. Leency advises 1 oz. of trisnitrate of bismuth, $\frac{1}{2}$ oz. of powdered catechu, 1 oz. of powdered chalk, 1 oz. of laudanum, and sufficient peppermint water to make 20 fluid ounces; give one teaspoonful to very young lambs, adding another for each fortnight); a bottle of laudanum (to be used carefully and sparingly); a bottle of castor oil (in cases of constipation); a good knife (curved inwards at the point); Glover's needles and thin tape (for use in cases of eversion), and such other appliances as the shepherd understands the use of (which are generally very few); some vinegar with blue vitriol in solution (to dress the feet of lambs or ewes which have become raw through being on wet litter, etc.); and a bottle of foot-rot mixture (to dress the ewes suffering from foot-rot, which should be attended to as early as possible).

Lambing, or Yeaning.

When the time of lambing approaches, the sheep must be closely watched, as it may be necessary to give the ewes assistance. Many of the most complicated cases of lambing occur among the first few, as those ewes which

lamb prematurely, carry dead lambs, or have other complications, are likely to come on early. The normal cases
come at normal seasons. If the ewe comes on naturally,
and the lamb comes right, it is generally best to leave all
to the course of nature. If the ewe has a small opening,
aid may be necessary, and this occasionally occurs in the
case of young ewes; but many troubles arise through
too early interference.

Signs of approaching parturition are generally noticeable

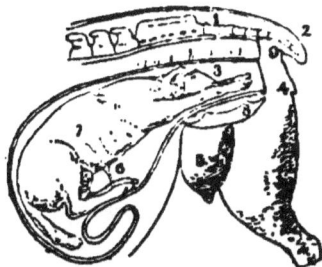

Fig. 10.—Section showing Natural Position of Lamb immediately
previous to Parturition.

1, Backbone; 2, Tail; 3, Passage; 4, Vulva, or Shape; 5, Udder; 6, Navel-
string, or Umbilical Cord; 7, Lamb Leaving the Womb; 8, Portion of Pelvic
Bones; 9, Anna, or Vent.

some little time before the lamb appears. For a few days
the milk forms in fair quantity, and the udder is somewhat
distended and red. Later the tail appears to rise up. This
is a delusion, as its prominence is due to the pelvic bones
having parted to allow room for the lamb to pass. Just
before lambing the water-bladder appears. If the lamb
does not come forward within an hour or two, there is
reason to think that something is amiss, though active
steps need not be taken for some time.

In the ordinary course the lamb will appear with its nose resting on its two fore feet, the hind legs drawn up under the body (Fig. 10) ; then little danger need be expected. If help is required, the legs should be drawn out singly; then stretching the opening with one hand to give the head a better chance of coming out, the body should be drawn forward by the legs, which should be pulled downwards towards the hocks of the ewe. The force should be applied at the same time as the ewe heaves. The lamb is covered with a thin skin (the placenta), forming a caul over the head, which should be removed from the nostrils to allow it to breathe. If the lamb is strong, it shakes and sneezes this out, but weak lambs occasionally have not sufficient strength to do so.

Malpresentations.

Sheep which have been subjected to high treatment and close folding are liable to greater difficulties at lambing than are those less highly developed and allowed to roam with little restriction. This is mostly shown by what are known as malpresentations, in which the lamb does not come forward in the normal manner. Malpresentations take a variety of forms, a few of the commonest of which may be explained.

The simplest form of malpresentation is where one or both fore legs are turned back, although the body is otherwise in a proper position. In this case the lamb must be pushed back until room can be found to bring the legs forward naturally. This should be done gently, and the advantage of a small hand is realized on these occasions. "There is plenty of room inside" is a good maxim for the shepherd to bear in mind in all cases of malpresentation. He should seek for the leg, get his fingers behind the knee, and gradually draw it forward; and if the lamb is lively, it may be well to slip a noose over its leg when brought forward, and then seek for the other leg. Having brought both forward, the lamb should be drawn out without delay, except to give the ewe a short rest if overcome by exertion. The force should not be too great. Some other irregularity, such as the head turned back, may not be noticed. Once, however, the legs and head are in line, the work of delivery may be proceeded with.

Sometimes the feet come all right, but the head may be turned back towards the shoulder; then it is necessary to partly replace the lamb, so that room may be found to turn the head into proper position. The slippery nature of the mucus covering the lamb makes it difficult to grip, and the head must be coaxed round by the fingers until a better hold can be got in the mouth or other convenient spot.

Occasionally a leg may have got over the head, causing an "arm-over-head" presentation. If the lamb is small and the ewe roomy, gentle movement of the legs may give relief; but generally it is necessary to shove the lamb back to gain the advantage of more room.

A simple case of one leg back often gives trouble, and in an ewe with little room, space will have to be found by shoving the lamb back far enough to permit the hand to get behind the shoulder.

The legs may be turned under from the knees; if so, they must be put into proper position. In fact, the legs are the main cause of difficulties, and the experience necessary to recognise when the legs are in place is of the greatest importance to the shepherd. To discern between the fore legs and the hind legs is a highly important matter, and to be sure of this and of their relative position is one of

the first requisites : for this reason the shepherd should, in
all cases when he finds it necessary to insert his hand,
try to impress on his mind the "feel" of the several
parts of the lamb. This is particularly necessary where
the lamb makes a rear or breech presentation with all the
legs forward (for the back legs are usually those most
forward) and the head turned under. If the fore legs can
be shoved inwards, the lamb may come away easily, rear
foremost; but it may be necessary to completely turn the
lamb to produce a normal presentation. A rear presentation with the legs backward (practically standing) may be
very awkward. Unaided delivery is almost an impossi-
bility, and the lamb must go forward and be delivered
either as a back presentation by getting the hind legs up,
or by turning and making a normal presentation.

In putting back a lamb, it is always advisable for the
ewe's hind quarters to be higher than the fore quarters.

Twins, in the majority of cases, come fairly well, and
are delivered with reasonable care, because they are not
of abnormal size. This is not, however, always the case,
as sometimes one lamb is unusually large and the other
very small. When the lambs are both making normal
presentations there is no particular trouble, unless both
come forward at once, when one must be shoved back and
the other taken out, leaving room for the second. They
sometimes lie "head to feet"; and one may be taken as
a normal and the other as a rear presentation if there is
room, or each must be treated as an individual case, according to its position. Care must always be taken not
to mix the legs of the two lambs when help has to be
given; and many cases where there are twins are
erroneously treated because one lamb only is assumed to
be present.

When the ewe's time has come, and the water-bladder
has appeared, yet no other signs are apparent, as the ewe
will not strain, or straining cannot bring the lamb forward, there is often a dead lamb within her, beyond the
reach of the shepherd. Artificial pains should be induced
by means of ewe drinks (sold specially prepared for the
purpose) in which there is generally ergot. These are
not always effective, and the ewe frequently succumbs.

Before lambing, ewes are liable to eversion of the uterus,
when the breeding bag is forced out. A simple and
useful clamp for retaining the bag (Fig. 11) is to be

Fig. 11.—CLAMP FOR EWES.

obtained from Mr. Huish, 8, Fisher Street, W.C. When
the bag is protruded it should be put back, and retained
by the clamp; or by a piece of tape stitched across the
shape. The tape should be wiped with an antiseptic
(such as carbolic acid and olive oil, in the proportion of
one to seven, or one of the advertised carbolised oils) to
prevent contagion of any kind. The tape must be cut,
or the clamp be removed, when lambing is imminent,
otherwise the ewe will be badly torn, or she may not be
able to get rid of her lamb, and die from exhaustion.

As soon as possible after being born the lamb should
be induced to suckle. Healthy lambs give little trouble,
as they soon find the teat. In the case of a weakly one,
the teat should be placed in its mouth and a small

quantity of milk be milked into it. It will soon gain strength. It is advisable in all cases to draw the teats to insure a clear passage for the milk, as sometimes they are blocked with dirt.

Whilst handling ewes, the shepherd should keep his hands well washed, or, going from a foul one, he may take disease to a healthy one. This is especially necessary in cases where there is a dead lamb.

If the sheep is long in lambing, or gets the skin or womb ruptured, or has produced a dead lamb, a good quantity of a mixture of carbolic acid and olive oil, in proportion of 1 to 7, should be injected, and the shepherd should rinse his hands in the mixture.

Dead lambs should be buried, as also should the after-birth of any lamb which is born prematurely. Dogs should never be allowed to eat dead lambs before they are skinned, as they are liable to acquire a taste to satisfy which they may become sheep worriers.

Prevention of Infectious and Contagious Diseases in the Lambing Pen.

Mr. Harold Leeney, in a very instructive paper on "The Lambing Pen," which appeared in the *Royal Agricultural Society's Journal* of March, 1897, says in respect to infection in the lambing pen :—" The bulk of losses at lambing time occur through want of systematic disinfection of hands, implements, appliances, and buildings. Nor is it any disproof of the efficacy of antiseptics to point to a record of success. The infectious elements are not always equally active, and we have to consider whether those ' unlucky ' or bad lambing seasons might not turn out a good deal better if attention were given to some of the simple precautions about to be mentioned. The remedies

and medicaments with which the shepherd should be provided are not many, but their constant use is of importance.
They include a bushel or two of lime, fresh burned and
ready for slaking, when it is desired to disinfect the
earth; a gallon of Jeyes' fluid, or other similar preparation that will readily emulsify in water; a quart or two
of carbolised oil, in proportion of 1½ oz. of carbolic acid
(pure) to each quart of olive oil; a cake or two of carbolic
soap—preferably Calvert's or some maker's whose guarantee
that it shall contain 15 per cent. of acid can be relied on—
the common soaps are variable and cannot be trusted;
half a dozen penny sponges; some soft clean rags, such as
old calico underclothing; and a bowl or metal pail (not a
wooden one) for washing hands, etc. These are the
essentials for disinfection and for antiseptic purposes."

Daily Work in the Lambing Fold.

The general procedure in the lambing pen can be best
explained by describing the shepherd's daily work in the
pen.

In the case of large flocks, the shepherd should not be
given other work to do during the busiest season of
lambing. He has to give attention to the ewes both night
and day, and get his sleep as best he can. If he does
not get reasonable rest some part of the work is neglected,
and a small amount of neglect may result in large losses.
We have seen several instances of this. Beyond an occasional look at other portions of the flock, he should not
be troubled with them, and if he is hard driven he should
have plenty of help with the ewes. The loss of an ewe
involves as great a loss as would pay a labourer's wages
for three or four weeks, and many lambs worth several
shillings each can be saved every year by proper attention.

The first thing for the shepherd to do in the morning is to attend to the lambs which have been born during the night, to see that they suckle; then the ewes about to lamb should be looked after. It is more convenient to put these in separate small pens around the yard, as they can then be found and handled without loss of time. Older lambs should then be looked to, and those which are weakly should be suckled or fed from a bottle. The ewes should then be fed with roots, and the backward ewes be taken out to their fold. Having had his own breakfast, if his attention is not required with an ewe about to lamb, he should litter the pens afresh. Foul pens are very productive of foot-rot, and, as it is not advisable to turn the sheep about too much whilst they are in-lamb, the feet are frequently in a bad condition. The ewes require their corn and fresh hay; lambs require attention, particularly those which are weak or where the supply of milk is short.

In case a mother dies, or cannot support her lamb, the lamb must be fed from a bottle or from ewes which have a surplus supply.

Where an ewe possessing plenty of milk loses her lamb, she should be given another. When the ewe and lamb are placed in a small pen, the difficulty of making her take to a fresh lamb can be overcome in a few days if patience is exercised. She must be tied up, otherwise at first the lamb will get no milk, and it is often necessary to stand beside her for a time to let the lamb suckle. The best means of restraining her is to drive two stakes firmly into the ground, and then to place her head between them, allowing room for her neck to move up and down, but keeping her head firmly secured between them. Room should be given on either side to allow the lamb to escape in case she tries to trample on it. If she is exceedingly savage it may be necessary to place another stake on either side of her, so that she cannot knock about. If the skin of her own lamb is placed on the foster lamb, or if the latter is rubbed with it, she will recognise the smell and will be more likely to take to it.

As soon as the ewe recovers from lambing she should be dressed if she has foot-rot, to prevent it from spreading, and the lamb's feet should be kept clean and healthy. At the least sign of lameness the lamb's feet should be cleaned and wiped dry, then a small quantity of a mild caustic (sufficient to moisten the abraded parts) should be placed between the claws to harden them and prevent foot-rot from being established. A mild caustic of three ounces of blue vitriol (sulphate of copper), dissolved in a pint of vinegar, effects this without inconveniencing the lamb, and is preferable to stronger caustics.

The sheep require feeding as night approaches, and should receive roots, hay, and the second portion of the corn. The backward ewes require bringing in for the night, and those which have already lambed sufficiently long to go out into the fields during the day, should be brought in also.

The shepherd has to prepare for a long night, as early in the season the days are short. The last thing to be done in the evening is to make sure that all the lambs are well fed, and special attention should be given to those which require suckling from their mothers or feeding from a strong-necked bottle. A short piece of elder, from which the pith has been extracted, inserted in the cork of the bottle, makes a convenient and safe mouthpiece.

Having got his sheep safely folded, the shepherd's attention is chiefly required among the ewes which are

likely to lamb, so that he may be at hand to give assistance where necessary. He must get his sleep as best he
can, and for this reason the advantage of living near to
the yard, or better still, in a temporary hut, is evident.
It may be necessary for him to be about nearly the whole
night when lambing is going on rapidly. At other times
it is sufficient if he looks round every two or three hours.
The advantage of having a large number of small pens
handy to turn in the ewes as they look like lambing is
especially marked at night, when the only assistance the
shepherd has in finding them is a lantern.

It is advisable to mark each single lamb with a spot of
ochre, and each twin with two dots, so that they may be
recognized. Confusion is often saved by doing this.

If ewes get low in condition, and weak through troubles
at the time of lambing, they should be given easily digested
and nutritive food. Concentrated strengthening powders,
specially prepared for the purpose, are valuable aids to
the shepherd in such cases, and should be kept in readiness. Strong, warm gruel of oatmeal, ginger, and strong
ale is serviceable when these are not at hand. The gruel
may be made of a pint of ale, one ounce of oatmeal, and
one ounce of powdered ginger.

LINCOLN RAM—TWO-SHEAR.

CHAPTER IX.

AFTER-MANAGEMENT OF EWES AND LAMBS.

In the Field — On Grass — Tailing and Castration — Older Lambs.

In the Field.

IT is a most important matter to get the ewes off the foul litter of the lambing pen as early as possible; therefore as soon as the lambs are strong enough, they should be taken out of the fold during the day-time, and be put on dry land. The lambing fold, as has been stated, is very likely to produce foot-rot and joint-ill, and these should be avoided by every possible means. Shelters, formed by hurdles stuffed with straw, should be set about the fields to protect the lambs from cold winds and wet, and should be arranged so that the lambs can take refuge from the wind whichever way it blows; or sheltering cloths (Fig. 9, p. 64) may be used for the same purpose.

The ewe's food should be liberally supplied, so that she may provide a good flow of milk. Among the foods which particularly conduce to the flow of milk are oats, peas, decorticated cotton cake, linseed cake, dried grains, with ensilage, grass and roots. Wheat and barley are also valuable, although some hold that the latter is not. It is best to give them in mixture, rather than singly, as the admixture acts beneficially. The lambs cannot digest food other than milk for a few weeks, during which, of course, they are dependent on their mothers. Maize, barley and turnips do not make a good mixture, as the nitrogenous constituents are lacking. Oats are especially valuable, but for some little time before weaning the quantity should be restricted, as from the extent to which they excite the flow it is difficult to dry the ewes as rapidly as is safe. All corn should be sweet and free from mustiness, as sour corn or cake makes the milk unwholesome.

From half a pound to a pound of a mixture of corn and cake should be given daily to the ewe, according to the supply of green food and the purpose to which it is desired to put the lambs. If they are to be fattened off quickly as fat lambs, the full quantity should be given whilst they "take their corn through their mothers"; afterwards it can be gradually withdrawn from the ewes and given direct to the lambs. If the lambs are to be kept on to be fattened off as tegs, the smaller quantity of corn will suffice. The lambs, however, always show a profitable return when the "lamb's flesh" is kept on them as long as possible.

The lambs should be induced to feed as early as possible; if a small quantity of finely ground linseed cake or pea husk is placed in a small trough outside the ewes' pen, and if a properly constructed lamb hurdle, such as will allow the lambs to pass through, but keep back the ewes, is placed near, they will soon find their way to it, and learn to feed. The lambs should always have an opportunity of feeding in front of the ewes, whether they are on grass or roots. Nothing is more suitable for lambs than the young tops of turnips, kohl-rabi, or kale and rape. The former are valuable early in the season, but get too old when they run to flower; rape follows, and kale is extremely useful throughout the spring, but more especially from March onwards, when the plants have thrown their spring sprouts. Failing young sprouts or fresh grass, mangels cut into fine slices are a good substitute. It is important that the lamb feed shall be fresh, and even in the case of corn or roots fed in troughs only small quantities should be given, so that the lambs may clear them out quickly. They will not return to stale food, and if any is left over it should be cleared out of the trough and given to the ewes, which are not so par-

Fig. 12.—SOUTHDOWN LAMBS, SHOWING EARLY MATURITY.

ticular. We have found nothing which keeps lambs in such healthy condition as the husk of peas—"pea chaff" as it is often called. Though not containing so much nutriment from an analytical point of view as some other foods, it appears to be easily digested, and the digestive organs are kept in vigorous condition by it.

It is astonishing what a large quantity of rich corn and cake lambs are capable of digesting when they receive their mother's milk. The one seems to aid the digestion

of the other, and they do not suffer from overdoing, as is
frequently the case in after life. With the improvement
in systems of feeding, and the greater aptitude to fatten,
sheep of the Down breeds particularly mature early. The
group of Southdown lambs in Fig. 12 is good evidence
of early maturity.

On Grass.

Less attention in the matter of feeding the ewe is generally practised as the lamb becomes more independent of
her. It is a mistake to let the ewe get too low in condition ; at the same time extravagant feeding at this period
is not warranted. In the management of a breeding flock
economy, consistent |with the health of the sheep, must be
exercised, or the cost of the lamb becomes excessive. An
ewe when drafted lean from the flock after breeding three
or four lambs, is worth no more—often less—than she
was when put in the flock as a theave. She spends
each year in producing a lamb, and the lamb's cost at the
time of birth is that at which the ewe has been kept
through the year, with slight additions for percentage of
loss and a proportionate share in the cost of the ram. On
grass this expense is slight, and on arable land it is dependent on the cost of raising crops.

A lamb ought to be born costing not more than ten
shillings, although it may pay in highly-bred show flocks
for it to cost twice that sum, a large portion of it going
in the cost of the ram. It is fair to charge the lamb
with the cost of the corn the ewe receives for two months
after lambing, as it is taken that the lamb is receiving
corn through its mother.

When the lamb gets its own living, however, the ewe
ought to be costing little more than at other seasons. The

food of the ewe should not cost more than twopence a week throughout the year, as other incidental charges bring up the total to ten shillings.

The advantage of getting a good percentage of twin lambs is easily apparent, as is the necessity of keeping as many as possible alive.

It does not pay to let ewes get feeble from lowness of condition, as some permanent injury is often brought about by so doing. Except, however, for a short period after lambing, the ewe must be regarded as the farm scavenger, often utilising that which it would be unprofitable to keep other sheep on. The first picking of the food should therefore be given to the lambs from the time they are strong enough to feed.

Tailing and Castration.

When lambs are about twelve days old the ram lambs not intended for breeding purposes should be castrated, as at that period they suffer very little from it. In some

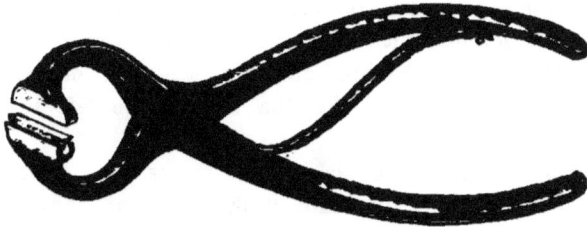

Fig. 13.—Lamb-castrator.

districts the operation is left until nearly weaning time, when emasculation, followed by searing, is adopted. The simplest method is to cut a slit in the side of the scrotum and press out the testicle, then to draw it out with pliers (Fig. 13) or by the teeth.

G

The operation of tailing the lambs should be done at the same early age, when both sexes are dealt with, and if the skin is pressed back towards the body, and the tail is cut through at a joint, the operation causes little pain, and the wound soon heals.

Castrating and tailing are best done on the evening of a fine day, and the sheep should be kept quiet during the night. With older sheep the wound is generally seared, but on young lambs a little tar to keep away flies is all that is necessary. Tailing is done to promote the comfort of the sheep subsequently. Breeds with much wool on the tail are liable to get very dirty tails when the animals, through change of food or other causes, scour. A long tail under such circumstances becomes a source of discomfort and ill-health. The foul matter collected about it is a great attraction to the flies which produce sheep maggots, and it is almost impossible to keep them away. Breeds which live almost exclusively on thin mountain pasture and produce wool of a loose and hairy nature are less liable to injury in this way, and their tails are frequently left uncut. The tail is the natural covering and protection of the hind-quarters, therefore on exposed hills it is advantageous for the tails to be left in their natural state.

Lambs intended to be sold as fat lambs must never lose their *lamb flesh*. By lamb flesh is meant that plump sleekness associated with the appearance of a young healthy lamb. This rapidly disappears when, through ill-health or shortness of food, the lamb becomes poor and pot-bellied, and once lost can never be wholly regained. It, however, may be kept up until the lamb is some three months old if proper feeding and care are bestowed on it. To some extent it may be continued longer, but unless

fed at high pressure the frame attains the looser appearance associated with young sheep. Although uncastrated
lambs grow faster for a time than do those that have been
castrated, yet castrated lambs make the best fat lamb on
the same amount of food; they handle better, and die
better. Lambs that are being got forward quickly for
the butcher should have the first pick of the choicest
feed, and their mothers should be liberally supplied with
corn.

Older Lambs.

A variety of green food is beneficial to lambs; in fact
they can hardly receive too varied a diet. The finest lamb
feed of any individual crop is white clover, but care has
to be exercised in its feeding, as in place of ordinary digestion it is liable to produce a large quantity of gas in the
stomach, causing what is known as *hoven* or *blown*. At
times the stomach is so distended that it bursts, and the
lamb quickly dies. The first sign of hoven is usually
frothing at the mouth. If this is detected no time should
be lost in getting the lambs out of the field; the exercise
of walking is beneficial, as it helps to move the gas.
Should a sheep become so bad that fears of its safety are
entertained, it should be relieved by the insertion of a
trochar and cannula, when, on the withdrawal of the
trochar, the gas will speedily escape. The cannula can
then be removed, and the wound will speedily heal. Special
care must be exercised for a few days to prevent recurrence. Failing the proper instruments, relief may be obtained by means of a stab by a knife. The hole made
may be kept open as long as desired by inserting a stout
quill. The exact position to insert the instrument is
equidistant from the hip-bone, the last rib, and the lateral

processes of the backbone, and the direction should be
nearly horizontal with the point directed slightly downwards, otherwise the kidneys may be injured ; nor should
it be done on the right side, as then other organs are en-
dangered. The point at which to thrust the instrument
is shown in the centre of the triangle in Fig. 13, and the
direction of the thrust is also indicated.

Clover, being very succulent, is the most likely green
food to produce hoven, especially when hoar-frost or moisture are on it. The best means to prevent this is to put

Fig. 14.—INSERTION OF TROCHAR AND CANNULA.

the sheep on the clover when their stomachs 'are well
filled, to prevent their eating too ravenously. Other foods
occasionally cause it, and it is not restricted to lambs, as
sheep of all ages may be affected with it, especially in
windy weather, though why windy weather should have
any influence is not apparent. The shepherd, however,
should always be on his guard.

As the autumn-sown catch crops come in, the lambs
require them, and if they can be changed two or three
times a day, so as to get variety, they thrive better.

Grass, or seeds, green tops of rape or kale, and vetches or other autumn-sown catch crop, make an excellent diet. When cabbages come in they are very valuable, as there is scarcely anything on which lambs thrive so well.

The great aim is to keep the lambs from diarrhœa or scour. They lose flesh rapidly when they scour, and it takes a long time to get them into a thrifty condition again. Sweet food is absolutely necessary, and nothing tends so much to rectify the bowels as a small quantity of winter barley, grown as a catch crop. It would pay every sheep farmer to grow a small piece of this crop every year, as it gives a good return, in addition to its medicinal properties. Another remedy for scour is to allow the lambs to nibble the shoots of a whitethorn hedge. The astringent properties of the shoots have a decidedly beneficial effect.

When the lambs are weaned they may be regarded as tegs. Until recent years it was not until autumn that this title was given them, but owing to their more rapid maturity under present systems of breeding and management, the career of a sheep that is intended to produce mutton is much shorter than it was. Except among the less improved breeds, existing under conditions which permit little alteration in feeding, as on hill land, three or four-year-old wethers have become a thing of the past. In the more highly-developed breeds the sheep pass through the comparative stages of lamb, teg, and wether, in as little as ten months. The lambs are sometimes shorn in July, and so lose their teg locks; the short wool which grows in the autumn gives them the appearance of wethers, and they go to the market as wethers.

CHAPTER X.

TEGS.

In Summer—Summer Keep or Feed—Concentrated Food
—In Autumn—In Winter.

Tegs in Summer.

REGARDING the lambs from weaning time as tegs, and speaking of them as such until they are a year and a quarter old, and in the natural course are shorn, the object is to get them fit for the butcher from November until spring. On grass-land farms they spend the greater portion of the time at grass, and it depends on the quality of the grass and the amount of corn they receive throughout their career, and on the hay and roots brought to them in winter, as to whether they go out as tegs, or are well on to be fattened out as wethers in the following summer. On farms where there is a large quantity of grass, but a fair proportion of arable where roots can be eaten off on the land in winter, they get a run on the stubbles in autumn, and probably white turnips, rape, or cabbages as a night fold in addition to pasture. In winter and early spring, until fresh keep appears, they are folded on roots, receiving hay and corn if not intended to be fattened out as spring tegs; and corn is less often given if they are only required to grow through winter and be fattened out when keep becomes more plentiful in summer. Good grass land should carry from ten to twelve

80

"OLD NEW" LEICESTER—TWO-SHEAR RAM.

Bred by Mr. Buckley, of Normanton Hill. Reproduced from coloured portrait by W. Nicholson, R.S.A., published in 1841.

sheep per acre from the time it is fit to stock in spring
until autumn, some being drafted out from time to time
as they fatten. On moderate pasture it is commonly
reckoned that an acre will carry two ewes and their offspring, which may be taken as five sheep in all.

Summer Keep or Feed.

The lambs, on weaning, should be dipped in one of the
well-known dipping solutions to free them from vermin
which may be on them, and to prevent others from locat-
ing themselves. This also has the effect of rendering
them less likely to be struck by the fly which lays its
eggs on the wool, the eggs speedily developing into maggots. The feet must be kept sound by paring, and if
disease breaks out they must be dressed with a caustic
foot-rot solution (sold for the purpose).

On farms where arable land greatly preponderates the
tegs are kept to a great extent on the arable land at all
times, as the grass is required for other purposes; though
if other kinds of keep are stale or short a fresh run on
the meadows is desirable.

During the early part of summer, when they are first
weaned, it is generally easy to find fresh, sweet food for
them, but as the season advances it becomes more difficult.
Unless lambs get sweet keep they soon "go wrong."
"Going wrong" conveys a special meaning to sheep-
keepers, as it implies that there is a derangement of the
digestive organs, which is shown by the lambs becoming
constipated or too open, more usually the latter; the wool
becomes dry and harsh, and instead of lying smoothly and
sleekly, is rough and broken, while they lose flesh, and no
amount of food appears to do them good. This state of
affairs is very commonly brought about by stale or sour

keep. Food may be freshly grown, succulent, and to all
appearances favourable for animal food, and yet be distinctly injurious to young sheep.

"Stale food" is most commonly found growing on land
which a short time previously was fed off by sheep. It
becomes soured by the droppings of the sheep, though in
what way is not definitely known, but it has been suggested that the souring is due to a larger quantity of
magnesia being taken up by the plants from the manure
recently deposited. More probably, however, it is caused
by parasitic worms. Rape fed off by sheep in summer is
soured; yet after standing over the winter, and being
subjected to frosting, it becomes sweet and wholesome, no
magnesia having been dispersed from it, though it may
have undergone some chemical change which has not been
discovered. Temporary leys in which there is a large
proportion of clover are dangerous in this respect. However,
if the young sheep are not turned on again until after a
crop has been taken off by mowing, no harm will as a
rule come of it. It appears that the first flush of growth
after feeding is the dangerous portion. Grass land is influenced to a lesser, though still a noticeable, degree.
Cabbages fed off by sheep and allowed to sprout again are
dangerous to young sheep if fed before winter. Old sheep,
having more robust systems, do not feel the effect in a
similar manner.

In feeding off his crops during the early part of the year
the farmer has to look forward and arrange that there will
be a supply of sweet fodder throughout summer. The
aftermath, seeds and meadows are a source of reliance, and
cabbages and kale are of great value during the period ·
between hay-time and the end of harvest, when there
will be fresh stubbles, and probably white turnips avail-

able also. Early rape may be fit to stock, though only
that sown very early will be sufficiently grown. Autumn-
sown vetches should be available until July, when spring-
sown vetches should come in. If a good succession of
crops like these is grown there should be little serious
difficulty in finding ample sweet food for the tegs until
autumn. Our own experience has shown us that there is
nothing equal to cabbages and kale as food from July to
September; they are less affected by drought than are
most crops, and, as it is not of so much importance that
they are fed at a definite period, they may be held over
for a time, or some other portion of the cropping may be
diverted to other purposes. Whatever provision is made,
however, it is imperative that the young sheep shall not
be forced to go on to stale keep. That various kinds of
worms have a direct influence on unsound or stale keep,
however, appears to be pretty certain, and attention is
directed to the discussion of the subject in Chapter XIV.

Concentrated Food.

Through summer the sheep receive cake or corn in accordance with the views the farmer holds as to the time
at which they shall go to the butcher. Those which are
to run the whole year round and be sold as wethers require none, though we like to give about half a pound
of pea-husk per day as a corrective. Ewe tegs to go into
the flock can be treated similarly. A quarter of a pound
of corn or cake per day is sufficient where the tegs are
not fattened out until spring. Where they are being fed
fast to be fattened out from November to January, from
half to three-quarters of a pound may be necessary.

Sheep do much better when they receive a fresh fold
daily, and they waste less food. A good fall-back, how-

ever, may be allowed, though it is generally better for the
ewes to clear up behind them than that the young sheep
should have to graze too closely.   This, however, is to
an extent dependent on the object in view.

Tegs in Autumn.

In autumn the food varies; the stubbles and young
seeds afford a fresh run. Some farmers object to allowing
sheep on young seeds, but, except on thin, weak plants, or
on heavy land in wet weather, when the sheep make deep
footprints which are left as cups to hold up water during
winter, the treading the land receives is beneficial to the
young plants, as the earth is more firmly pressed round
the roots, and there is no need to graze so hard that the
heart of the grasses or clovers is gnawed out. The ordinary practice of rolling does not do nearly so much good,
because the soil is not so regularly tightened about the
roots.   The Drumhead cabbage, white turnips, and other
crops come into season now.

The winter is the season for feeding hard roots, such as
turnips, swedes, and kohl-rabi. It is important, as far as
possible, to get the sheep gradually broken to turnips, and
not to take them straight from grass, seeds, or other foods,
and put them on roots. This sudden change of diet often
causes great loss of life, while many sheep receive a check
in growth and fattening from which they do not speedily
recover. So well is this recognised that in buying-in in
autumn, those sheep known to be broken-in to roots are
worth a shilling to two shillings a head more than are
sheep in every way similar except that they have received
no roots, but have come straight from grass. Cabbages
and rape are good stepping-stones to roots, and white tur-

nips (probably owing to their ripeness) are less injurious
than swedes.

It is advisable to give very small quantities of swedes
at first, and to give with them a plentiful supply of dry
food, such as hay or hay chaff. The importance of dry
food throughout winter cannot be too vividly remembered;
nor should the season be allowed to advance too far before
it is given. In the case of either fattening or breeding
sheep, a plentiful supply of dry food is the most effective
means of preventing losses in winter.

Tegs in Winter.

When the tegs get on to roots the turnip-cutter
(Fig. 15) should be set to work. Cabbages, kale, rape

Fig. 15.—TURNIP-CUTTER.

and white turnips are soft enough to gnaw, but swedes
and kohl-rabi should be sliced. In the case of sheep
which it is desired to get out very quickly, it is preferable to slice white turnips, as they satisfy their appetites
more easily, and have thus a longer time to rest and

digest their food. The saving in food repays the cost of labour, and the sheep thrive better.

The corn should be increased as the sheep become fatter and it is desired to get them quickly to the butcher. Under ordinary circumstances a pound weight per day is all that it is advisable to give. Larger quantities are sometimes given, but the risk of loss from overdoing is great. "There is another sheep dead this morning, sir," is the unwelcome news every farmer who fattens large quantities of sheep at high pressure knows only too well. It occasionally results from eating too many unripe roots, but generally from what is known as "making too much blood." It is, indeed, a form of paralysis caused by too much nitrogenous matter in the blood, this being brought about by a diet containing an excess of nitrogen. This presses on the brain, and the sheep rapidly succumbs if assistance is not given. The obvious remedy is to weaken the blood, and the shepherd does this by drawing off a quantity as soon as possible. The corn must be withheld, and only gradually returned to the sheep, as when one sheep shows signs of excess it is probable that others of the flock are in danger. It indicates that the corn is too nitrogenous, and that this must be changed for a mixture which is more starchy. A sheep with paralysis appears listless, and lies in a helpless condition. As soon as signs of this are apparent no time should be lost in relieving it of blood. Half a pint may be taken away with safety. The best vein to open is that on the side of the face, a little below the junction of the eye orbit and nose. Some bleed by cutting the ear; others from the vein on the inside of the leg, a little above the knee. If left too long the sheep will soon die. However, a careful shepherd never allows a sheep to die a natural death; when he sees that there is no chance for it he kills it and dresses it, so as to make it marketable. The advantage of a well-balanced mixture of corn is easily recognised when the danger from one in which too much nitrogen is present is understood.

All roots should be sliced for fattening sheep when once they get on to the harder kinds. It is best to eat the swedes in mid-winter, and to follow these with kohl-rabi after the turn of the year. No farmer should allow the first week of December to pass without having a portion of the earliest and ripest swedes—sufficient to supply the sheep for a month—safely clamped. Neglect to do this is the source of great waste during frost; the roots are more difficult and more expensive to get up when frozen in the ground, and are less valuable as food owing to their excessively cold condition; while in very severe frosts they are totally destroyed, and great difficulty is experienced in getting a supply of food during the remainder of the winter season. As a rule turnips may be pulled by hand. If very tightly in the ground a turnip-pecker (Fig. 16) is

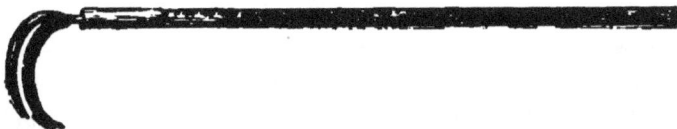

Fig. 16.—Turnip-pecker.

necessary. This is also needed where sheep gnaw the roots, as if the cups are not pecked up they are wasted. The hardy variety of kohl-rabi, particularly if not over-ripe, will withstand all but the most severe frosts, and need not be clamped. If ripe the plants may be got up and clamped in the same way as swedes. If the tegs have

been liberally treated throughout the season they should
be fattened off and sold from time to time as they become fit for the butcher, so that they are sold out by the
end of March.

Those which have to be run on and sold as wethers
should give little cause for anxiety; they are not fed at
high pressure, but receive sufficient food to keep them in
a healthy growing condition, when little risk is run.

The distinction between the fattening-out of tegs and
wethers lies chiefly in this: tegs receive corn practically
throughout their career, and wethers only at the end.
Teg-fattening involves considerable outlay in the purchase
of feeding stuffs; wethers are fed almost entirely on the
produce of the farm. A portion of the tegs are often kept
for selling out between March and June; that is, the last
of them are finished off on the first growth of grass or
seeds. These, as a rule, are not fed fast during the
autumn, and probably receive no corn until Christmas,
when they are given a small quantity; and this is increased slightly from time to time as they appear to need
it, a much larger quantity being given for the last month
before they are turned out fat. These tegs are often shorn
before being sent to market.

Tegs kept through winter on grass are generally maintained at little cost. There are instances where it would
pay to give more food, particularly in those cases where
sheep are sent out to grass on agistment to distant farms.
The fear that the corn supplied or paid for may not be
given to the sheep, probably restrains many from providing it; but, considering the loss of life, and the fact that
many of the sheep weigh less after the winter grazing than
when they are sent to it, something more is obviously
required. Where sent out to winter on roots some dry food,

such as hay, ought to be provided : otherwise the progress is very slight in relation to the money spent on their maintenance. The small additional expense makes a large difference in the growth and condition of the sheep. It is a foolish policy to spend ten shillings on wintering sheep in such a manner that they do not improve during that time, when half a truss of hay, at 1s. 6d. per truss, will supply a quarter pound of hay per day for nearly four months, and make a marked difference in their condition: half a pound per day is, of course, better.

CHAPTER XI.

WETHERS.

Shearing—First Feeding of Leys.

Shearing.

WETHERS supply a large portion of the summer and autumn mutton. Fat lambs are in demand throughout hot weather, but the greater portion of the sheep killed are wethers, largely composed of those which are bred in the later lambing districts, particularly the grass districts. The summer management of these is simple; they are stronger in constitution than are tegs, and they therefore are less inclined to be affected by sour keep or excessive feeding.

The shearing of the wool turns the teg into a wether. In the south-east of England the term teg is retained after shearing; but this is the only exception.

Before shearing it is customary to wash the sheep to relieve the wool of the greater part of the dirt, and to give it a more marketable appearance. Of recent years a number of advocates for selling the wool in an unwashed condition have sprung up, and they have found followers. It is urged that it is to the advantage of the farmer, who, though he gets less per pound, is more than compensated by the greater gross weight paid on. This is opposed to all other experiences. The buyer knows more about wool than a farmer with a limited experience is likely to. The

SHEEP-WASHING FROM THE BANKS—AN IMPROVISED WASH-HOLE.

object in offering goods for sale is to make them appear
more marketable. Dirty potatoes, uncleaned wheat, un-
groomed horses—in fact, any commodity which is not
offered for sale in an attractive form—will not realize so
much as when properly prepared for sale. Why, then,
should wool? If a farmer can find an inexperienced buyer
who can be misled as to the quantity of dirt in and on
the wool, he may gain an advantage, but such buyers are
not commonly met with. If there is a doubt as to the
actual condition of any article, and if the professional
buyer has reason to believe he is running risks in buying,
he does not buy confidently, but buys within the actual
value, so that he may cover his risk, and thus the seller
is the loser. It is to the advantage of the wool-buyer to
purchase wool in a crude condition because his special
knowledge places him in a better position than the seller
with less experience can hope to be. He will not give
more than it is worth, but will probably pay less.

Sheep washing is performed in various ways, though
most commonly in running streams, the sheep being thrown
into it to soak for a few minutes, and then well rubbed
with long-handled scrubbers, the operator standing on the
bank. An illustration of this method is shown in the Plate:
advantage of a bridge over a small brook was taken to
dam back the water, and a landing place cut in the bank
a few yards up stream, a washing place being thus very
quickly formed. When large numbers of sheep are to be
washed it is desirable to give the washing hole a more
permanent form. It is necessary that the water be deep
enough to allow the sheep to be thoroughly immersed.
The sheep should be dropped in rump first, as the water
thus breaks through the wool more easily. Particular
attention should be paid to the back about the loin, as

H

most dirt accumulates there. The most effective sheep
washing place is found where water can be brought from
a higher level by means of a trough, as in the case of
the overshot water-wheel, as the falling water forces out
the dirt more thoroughly, and less scrubbing is required.
In such cases it is common for the shepherd to have a
tub sunk convenient to the spout, in which he can stand
dry, so that he can hold the sheep in the required position.
Occasionally sheep are washed in tubs of hot water, but
this is generally practised in warmer climates than that
of England.

At the same time, it is a mistake to shear sheep after
washing until the "yelk" is up again. Here the buyer
has a point. The natural yelk, an excretion from the
skin, is beneficial in the subsequent treatment of the wool:
this is washed out, and it does not rise for a few days
after washing; the time depends on the temperature, it
being quicker in warm than in cold weather. The farmer
loses the weight of the yelk if he does not allow it time
to rise.

Washing renders shearing easier, and the sheep is more
marketable than when shorn badly, as it must be when
its back is full of dirt, for the shears will not keep a good
edge, and they will not face the dirt. In some districts,
however, shearing is so badly executed that, dirt or no
dirt, the sheep are roughly turned out. Taking everything
into consideration, we prefer washing before shearing.

Sheep shearing is performed by the ordinary shears, or
by the mechanically driven shears made on the principle
of horse clippers (Fig. 17), actuated by any convenient
power (Fig. 18), such as steam, horse, or water. There is
considerable diversity in the manner of shearing by hand.
In some districts the shearer mainly stands whilst working:

in others he kneels, the sheep in these two cases resting
on the ground; in others the sheep is laid on a bench, or
in a cradle, whilst the shearer stands. In some districts

Fig. 17.—MECHANICAL SHEEP SHEARER.

the shearer cuts longitudinally from end to end of the body :
in others he cuts right round, starting at the belly and

Fig. 18.—POWER MACHINE.

going over the back down to the belly on the other side :
in others the cutting is done against the natural fall or

hang of the wool. The latter makes the best work. The
Plate shows three representative attitudes in shearing: (1)
opening round the neck ; (2) the middle of the body; (3)
the hind quarters. The sheep should be perfectly dry at
shearing, or the wool will not be safe to store. All straw,
twigs, etc., should be carefully picked out of the wool,
and dirty locks be trimmed off. The opening up of the
work is performed in slightly differing ways, but the wool
about the head should be trimmed off first; some shearers
continue to shear round the neck down to the brisket,
merely laying back the wool with the hands: others
make a cut up from the brisket to the head to afford an
opening.   First-class work requires the shearer to be
equally dexterous with either hand, so that he may at
all times meet the fall or natural hang of the wool.
The difference in shearing with or against the fall of the
wool is as marked as that of cutting a laid field of corn
with and against the direction in which it lies. A man
who can work with but one hand must work one side
with the fall instead of against it. The wool hangs
vertically from the spine, except on the belly, where it
tends towards the opposite direction ; thus in shearing the
near side, to obtain the best results, the work has to be done
with the right hand from the belly to the spine ; and on
the underline with the left hand from the line of the navel
to the outside of the underline. In ordinary work, the
belly being out of sight, the wool is generally cut off
without regard to this. It is usual to shear all round
the neck to the shoulders, after which one side is com-
pleted before the other is commenced. A novice finds
considerable difficulty in holding the sheep conveniently.
The sheep should at all times be held in such a way that
the portion being clipped stands out prominently, as then

SHEEP-SHEARING—SHOWING THE THREE POSITIONS.

the skin is taut; whereas when loose it lies in wrinkles, which are likely to be cut. The skin may be further tightened by stretching the shorn portion towards the shearer; but if the uncut wool is pushed back small pieces of skin are lifted up and are inevitably cut off by the shears. The shearer learns by practice how to make portions he is working on bulge out, using his own legs and body for the purpose. In holding the sheep he often has to hold a portion of it between his arms and body, leaving both hands free to manipulate the skin and the shears. The cutting should not be done with the point of the shears, but with the portion two inches or more below. If done with the point there is great liability to rase the skin, causing it to show a red mark, and rendering it very easy of attack by flies. The shears should always be kept flat on the part to be shorn, with the points slightly raised. The lower blade should be employed as a guide through the wool to open it up and to make a definite shearing line; for this reason it must be kept steady and the upper blade be brought down to it. This distinguishes shearing from clipping. Clipping off locks is of a similar nature to clipping a hedge, where both blades are moved. A clipping action is distinctly opposed to low cutting, as the shears very soon "ride out" of the wool. The point of the shears should be thrust into the wool, and then the hand be allowed to fall back slightly; this permits locks which have been over-pulled so as to lift small pieces of skin into a dangerous position, to tighten on the body so that, being then in a natural position, they are not cut. The operation is unconsciously performed after a little practice; and it is of the greatest importance where the wool is matted together and where the skin is thinnest.

The shearing should be done in such a manner that the

lines are perfectly vertical, as then the neatest appearance is obtained; and it is highly essential that the wool be well cleared off the back along the spine. If the wool forms a sort of ridge along the spine the sheep's back appears narrow. Every one knows the advantage of a flat back as representing good condition. Very skilled shearers can make the left and right hand cuts meet exactly on the spine. .

The beginner should not aim at taking a wide cut, and the neatest work is done with a "blow" from half to three-quarters of an inch in width. Should the skin be cut, a small quantity of tar should be put on the wound to keep off flies, and to keep out injurious matter, for occasionally blood poisoning is caused when the wounds are left unprotected.

Care should be taken to keep the fleece clean and un-broken. It should be carefully wound into a bundle. It should be laid out flat, the inside downwards, then the sides turned over so that it forms a narrow strip eighteen inches or so wide; beginning at the tail end it should be compactly rolled up, and by twisting a portion of the wool about the neck to form a bend be tightly bound. Wool should be stored in wool sheets in a dry, sweet room, and it is essential that it be perfectly dry when stored. A damp floor or damp walls must be avoided.

When shearing stock sheep less care is necessary than when fat sheep are to be sent to the market at once, when the greatest neatness is desirable. In such cases the shearers take as broad blows with the shears as the farmer will let them; and the sheep present a very rough appearance at times. In taking a broad blow more wool is left on, and as in such work regard is not paid to the hang of the wool, it is not cut so short as it might be.

In Scotland a considerable portion of the shearing is done while the men are on their knees, only the opening up about the neck being done while standing. This is suitable for small mountain sheep, but the strong, heavy sheep in England are more conveniently held in the manner already described.

First Feeding of Leys.

Good grass pastures and temporary leys of grass, clover, sainfoin, and lucerne, afford good food for wethers, and they can be kept very economically thereon. While food is plentiful, particularly on the first feed-over, the sheep thrive with very little corn. When I was managing the Woburn experimental farm for the R.A.S.E., I had a striking illustration of the value of the first feeding of seeds as compared with subsequent feedings of the same crops. For the sake of experiment the crop was divided into plots an acre in extent, and the sheep received cake in one instance, maize in another, and on two acres no extra food. During the five years over which the experiment was tried in different places, those which received no corn increased as much as those which had it; but in the second feeding-off each year, although the food looked as fresh as during the first feeding, and the crop was as heavy, those which received corn went far ahead, yet to all appearance the clover was as good on one plot as on the others. As far as the feeding went the cake and corn were wasted in the first feeding. Nothing different from common practice was done, so it is highly probable that this has been the case in feeding off much similar cropping in other places. Close attention to the "doing" of the animals would probably show that much corn is wasted on old sheep at this time. Sheep can only make use of a certain amount of nutriment in whatever form that nutriment is supplied; and if the crop supplies all that is necessary it is obviously waste to give them anything additional. When the clover is stale, although luxuriant, as in the second feeding-off, the advantage of corn is at once apparent. Those who have a convenient weighbridge might easily put the matter to the test. A skilled sheep-keeper and feeder knows by handling whether the sheep are thriving satisfactorily; but every sheep-keeper, no matter how long his experience, is not necessarily skilled, although he may have a good general knowledge.

OLD WILTSHIRE RAM AND EWE.

Bred by Mr. Turner, of Hindon, Wilts. Reproduced from coloured portrait by W. Nicholson, R.S.A., published in 1841.

CHAPTER XII.

DAILY MANAGEMENT OF LAMBS AND TEGS.

*While with their Dams—In Summer—In Autumn—In Winter
—Fold-Setting—Unhealthy Foods—Selling out Tegs.*

While with their Dams.

HAVING given an outline of the management of
the lambs and tegs throughout the year, the details
of the work among them from day to day may be dealt
with.

As soon as the lambs get strong on their legs, the ewes
go out to feed on grass or roots, taking their young with
them. The lambs eat nothing at first, and should be
provided with shelter in the form of hurdles, set up to
break the wind. While very young, if the weather is
wet, it is well to make a covering shelter against rain,
similar to that in the lambing yard, as lambs, though
little injured by dry cold, are easily chilled by wet. In
the course of two or three weeks they will begin to nibble
at green food, therefore they should be allowed to run
forward on the freshest and tenderest keep; at the same
time they should be tempted to eat corn. Nothing is
better than pea husk, with a small quantity of linseed
cake broken finely; the dust from the broken cake should
be sifted out for this purpose. Very little of this should
be placed in a small trough in the lamb pen, near to the

lamb hurdle or gate provided for their exit. They should be taken back to the fold every night.

When they get stronger, the quantity of corn should be increased, a small quantity being given whenever they have cleared up the previous supply, but no stale corn should be left in the troughs. A small quantity of hay should also be allowed. It is more economical to feed hay from sheep racks (Fig. 19) than to let them eat it on

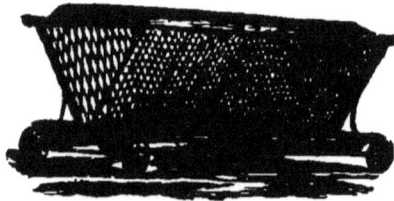

Fig. 19.—SHEEP RACK FOR HAY AND CORN.

the ground, when a portion is sure to be trodden in and wasted.

By the time they are two months old, they require feeding systematically, as they have by this time acquired large appetites, and must be regarded as little sheep, still receiving a share of their support from their mothers, but becoming less dependent upon the latter.

Half the corn at first given to the ewes may, by this time, be given to the lambs, and it may gradually be withdrawn until the ewes receive none. In seasons of short keep, it may be necessary to give both ewes and lambs corn to eke out the food.

If carefully fed, and no special disease breaks out, no serious trouble need be apprehended. The lambs require, however, to be kept free from lameness, therefore the feet should .be kept sound and in good shape. Too strong caustic frequently contorts the feet, and renders the animal more likely to contract lameness at all times through its life. The paring of the feet should be restricted to keeping them in a natural shape, and as the hoof grows quickly, unless the ground is dry, there is great likelihood of its not being worn down sufficiently fast to keep it in shape. If the lambs are caught once a week, and any abnormal growths are cut back, they will rarely suffer. However, in wet weather, especially if the lambs are kept on the dung in the lambing pen, there is likelihood of the skin between the claws being abraded, and lameness, at first from soreness, and subsequently from foot-rot, will develop. We repeat, a mild caustic, such as the solution of sulphate of copper and vinegar mentioned on p. 75, should be used. If this is done at the first appearance of lameness, foot-rot will rarely establish itself.

On grass, lambs get little help in the shape of green food beyond what they get from the pasture, and if the pasture is good they have a good diet. It is, however, beneficial for them to have other food, and if kale sprouts or finely-sliced mangolds are brought to them daily, they will profit thereby. Later, the autumn-sown catch crops are available, and a short time each day spent on these will be advantageous, corn, of course, being given as before mentioned.

The early maturity of Hampshire lambs is in no small way induced by the frequent changes of food supplied to them during a day, and this is evidence that lambs require a mixture to thrive to their utmost. Hampshire lambs change their feeding ground as much as four or five times a day on those farms where ram breeding is a feature, and growth and condition are essential to the success of their sale. Folds are set on vetches, cabbages, in the water meadows, on rape and temporary pasture, in which the lambs get the first run over, spending a few hours in each daily. This system is the most thorough practised, and, of course, need not be followed in its entirety where the most rapid feeding is not required. Under ordinary circumstances, lambs at three months old do well on two changes a day, and they do not then require to go to the lambing fold for shelter. If vetches, trifolium, or other catch crops are available, it is an advantage for them to spend a few hours there, whether they are on grass or on temporary pasture during the remainder of the day, though, of course, many do well folded entirely on good clover, to which a small quantity of other green food is taken. But it is undoubtedly an advantage to have some change in diet.

As the ewes' milk falls short, it is important that the lambs be supplied with water. If this is kept constantly by them they never drink to excess, though when given occasionally there is danger of it. A block of rock-salt should be kept in the fold, so that they may lick it as Nature dictates.

It is always recognised that lambs, in fact all sheep, do best where not in too large a number. This is particularly the case where lambs are receiving corn or special food, as the stronger shove aside the weaker. The weaker should, therefore, be kept by themselves where practicable.

In Summer.

At weaning the lambs should be taken a long distance from their mothers, so that they will not hear the call of the latter. Their food should be particularly good on weaning, so that they may not miss the effect of their mothers' milk. Their folds should be changed daily, so that they get a fresh feed daily. The dipping should not be delayed, as the wool is long enough to carry sufficient of the solution to keep them free from ticks, lice, and scab through summer. As they are small and have little wool, the operation is quickly effected. An autumn dipping is also beneficial, and the lambs will be more comfortable and thrive better for it in the spring. Properly constructed dipping troughs are supplied by makers of the better kinds of dipping powder, and these are more convenient than a rough arrangement of tubs. It is always advisable to allow the sheep to drain on a piece of fallow ground, so that the drippings of the solution do not affect pasturage which will be eaten. The sheep may be fed through the hurdles meantime.

The dipping has a very marked effect in keeping off the fly which causes maggot on sheep; but after a time it cannot be relied upon. Throughout summer one of the shepherd's first duties is to detect sheep which have been struck with the fly. The fly generally settles on moist parts about the tail, though the loin and shoulder are favourite positions. A keen shepherd will notice the discoloration of the wool very early, while a careless man will let the maggots commit serious pain and injury before he can see that the animal has been struck. Early signs are that the sheep is inclined to draw away from its companions, and that it switches its tail frequently. Later, it looks round at the place, and perhaps pulls the wool. Then the wool becomes loose and broken, and a streak of brown, foul moisture is seen. In large fields, with overgrown fences, sheep are occasionally lost in the surrounding ditches, through the carelessness of the shepherd, and are eaten alive.

Mercurial ointments were commonly used a few years ago to get rid of the maggots, but they were dangerous, as, if too freely used on broken skin, they were liable to cause poisoning. Simple remedies are found in milk and turpentine, and many other mixtures; but McDougall's Fly Oil is by far the best we have used, for it not only at once kills the maggots, but heals the wound. Before disturbing the colony of maggots, it is well to put a little of the mixture round it, to prevent their crawling outside the point to be dressed. If in a position where a blemish will not show, it is well to cut off a little wool above, then to pour in some of the mixture, and shake out the maggots. It is advisable to sprinkle a little flowers of sulphur on the wool about the place, to prevent any fresh attack.

If lambs scour they are liable to be struck about the tail; to prevent attack, sulphur should be sprinkled on with a dredger. Trouble is saved if the wool about the tails of all the sheep is trimmed to prevent the accumulation of dirt down the thighs. Nothing is more unseemly at any period of the year than filthy locks from the tail to the hocks. An old pair of shears should always be at hand to keep these parts clean.

The shepherd has a few special opportunities of discovering if any of the sheep under his care are out of health. These are when he first arrives in the morning, and when he gives them fresh food. His first duty is to look out to see if any of the sheep are at a distance from their companions, as this is a suspicious symptom. Such sheep should be closely noticed. Again, when fresh food is placed before them, whether in a new pen or in the trough, those which are failing linger behind. It may be taken that whenever a sheep does not take to its food there is something seriously amiss with it, and that it will soon recover or soon die.

Sheep medicines are few, as little scientific skill has, until recently, been devoted to their ailments. Under most circumstances little medicine is used, but among the best general medicines we have personal experience of Pettifer's Herbal Tonic, and Day & Hewitt's drinks rank highest, and should be available.

There is no excuse for a shepherd who allows a sheep to die, when he has once noticed any ailment. A sheep which dies a natural death is not worth a shilling more than its skin, whereas one which is killed and allowed to bleed is generally worth at least two-thirds its ordinary killing value. A shepherd should always possess a good knife with a long blade, suitable for cutting a sheep's throat, bleeding a sheep, and for paring feet.

It is important that troughs in which corn is given to sheep shall be moved daily, so that the sheep will not unduly manure one portion of a field to the detriment of others. This refers to pasture as well as arable land, and to all seasons. It is of highest importance in winter time, when feeding off roots which will be followed by a barley crop. It is necessary that the whole crop of barley shall be uniform, and if the sheep are brought to one place with undue frequency this place will produce a rank growth, which, in all probability, will be storm-broken, or at any rate, produce corn which will cause the sample of grain to be uneven. The folds should, therefore, be set regularly, and similar quantities of food be fed in each.

The tegs which are shorn in summer time should be

shorn in time to allow the fleece to grow again before winter, or they will not do well, as the wet will go through to the skin. Under natural conditions the wool rarely gets wet down to the skin, as it lies in a manner which conducts the rain outwards. If this wool is very short in winter the water runs straight through it, and the sheep, being chilled, cannot thrive. Ewe tegs which are not to be sold fat, but are to go into the flock, are not often shorn. Shepherding is easier in summer when tegs are shorn, as fly are easily detected, and clagging (trimming off foul and dirty locks) is not necessary.

In Autumn.

In autumn, when the sheep come on to the roots, the system of shepherding alters, as there is more close folding. During the day the sheep may have a run on the stubbles and young seeds, or grass, but at night they preferably go on to cabbages or white turnips. The folds should be made sufficiently large for one day's feeding, but a fair amount of room should be left for them to fall back upon, as in wet weather the pens get muddy, and on most land afford bad lair; the food also gets muddy. It may be taken as sound practice in autumn and winter shepherding that a good large fall-back is necessary, and that in summer it is advisable.

During the early part of autumn the turnip cutter is not required, as the food is soft; but later on, when the swedes are used, it is undoubtedly an advantage to cut the roots. The extra expense of cutting involves the necessity of getting up the swedes and cleaning them; however, the whole of the crop is eaten, and the sheep thrive far better. Against this the cutting is saved; but a portion of the turnip has to be pecked up or it is lost.

It at any rate gets dirty, and even ewes are damaged, as gnawing causes the teeth to get broken before they would otherwise be.

A shepherd should be able to attend to 200 fattening sheep if the turnips are got up into heaps made a chain square apart. There are thus ten heaps on an acre, each heap containing two tons. Allowing one cwt. per sheep per week, a heap would carrry 280 sheep one day; or 200 bigger sheep eating 22lb. per day. The turnips are thrown together as pulled, except that if they are to be stored for some time before being used the tops are cut off. The shepherd should clean these, cut them, and feed 200 sheep per day. In addition he should move and re-set the hurdles as wanted, fetch the cake, corn, chaff, and hay, keep the sheep's feet sound, remove boulders of dirt from their bodies, and keep them clean behind. It is also expected of him that he will dig out the patches of couch or twitch in front of the sheep. When no cutting is done he should be able to shepherd 400 sheep. Unless the shepherd has sufficient hurdles his work is very much cramped, as he may not be able to leave a proper fall-back, or he may not have sufficient to make a fresh pen before breaking up the old one.

One of the shepherd's chief troubles is to *keep the sheep within bounds.* On open plains and heath he relies very much on his dog during the day, bringing the sheep to the fold at night. On mountains, where the range is extensive, they roam very much at will, being overlooked in a general way, but shepherding as understood in enclosed districts is little followed. Almost all kinds of fencing are used to keep them in check, hurdles, either wattled or slatted, being most commonly used for close folding on roots, where by their close confinement they are most likely to exert themselves with the view of getting out. Wire-netting and string-netting (Fig. 20), although used for close folding, are more suitable for dividing fields and other large areas ; nevertheless, in districts where hurdles are difficult to obtain, they are of great service. Wattled hurdles are specially useful in exposed situations, as they break the force of the wind, and afford some shelter from the sun in hot weather.

Fig. 20.—SHEEP NETTING.

Sheep are liable to lie too much under them, and thus make an uneven manuring. Sheep are bad friends to growing hedges, as they eat the young shoots at the bottom, and thus weaken it in its most important part.

Hay is often fed to sheep through hurdles. This puts an extra strain on the hurdles, and is wasteful of hay. Hay-racks should be used. The simplest are those used in Wilts and Hants ; they consist of a narrow longitudinal frame into which spars are fixed from side to side so as to give a trough-like appearance, as shown in Fig. 21. These are filled with hay, and then turned over on the flat side, when all the hay is eaten without a portion

Fig. 21.—WILTSHIRE HAY RACK.

being trampled into the soil. Heavy racks are inconvenient to move in the field, while these are easy to move and simple to mend.

Wooden troughs are preferable to iron on account of the weight, while zinc and other metal troughs soon become battered in the rough usage they get. The simplest form is the ordinary pig trough or V shape, though those made with a bottom board are most capacious, and are better suited for containing chaff and corn as well as roots. An advantage of metal troughs made with semi-circular bottoms is that they are very easily cleaned out. The troughs should be 7ft. or 8ft. in length.

In Winter.

When the sheep are entirely fed on roots they require sufficient troughs (Fig. 22) for all to feed at one time. First thing in the morning the shepherd should clean out

Fig. 22.—WOODEN CORN AND ROOT TROUGH.

the troughs, and give the sheep their chaff and corn. If
the sheep have hay instead of chaff it can be placed in
the hay racks. If chaff is given they are induced to eat
more of it if they receive meal well mixed with it, which
is often desirable, particularly when first attempting to
make them eat a considerable quantity of dry food.
When giving the corn the shepherd should make his dog
keep back the sheep until all the corn is in the troughs,
so that all may receive an equal share. He should carefully watch them to see that they come up promptly,
and that when they have got to the troughs they feed,
otherwise he may miss seeing an ailing sheep. As soon
as the corn is cleared up he should fill the troughs with
cut roots until he notices they begin to draw away to
rest, and digest their meal. He may then get his own
breakfast. After this he may clean another supply, or set
another pen, according to necessity. Before turning the
sheep into a fresh pen he should go over it with a four-
tine fork, and dig out any small pieces of twitch which
may be there.

If any sheep show signs of lameness they should be
caught and dressed. Very often lameness is caused by
dirt, or stalks of turnip leaf, which should be taken out,
and a mild caustic (see p. 75) wiped in between the claws.
This will, in most instances, prevent foot-rot. In case of
foot-rot the sheep should be isolated, and kept from other
sheep until all traces of disease are destroyed. Whenever
a sheep is dressed for lameness a temporary mark of
ochre should be put on it so that it may be recognised
easily.

Before going to his dinner the shepherd should give
the sheep another feed of roots. A good opportunity is
generally afforded about dinner-time for the shepherd to

put up the cake, hay, and chaff that he will require for
the following day. In the afternoon he can continue his
work of cleaning a fresh supply of roots, moving hurdles,
or other necessary work ; and can give his sheep their
second supply of corn about an hour before he will leave
the pen at night. When the corn is cleared up he can
commence to give the sheep their supper. He should
continue this until all the sheep are satisfied, then he
should fill up the troughs and leave them filled. The
sheep will then rest contentedly through the night.

The last duty before going away is to carefully look
round the sheep, and notice that none are ailing. Then
all the hurdles should be tested to see that they and the
stakes are safely secured. After a drought, when it is
difficult to drive the stakes far into the ground, the
change to heavy rains occasions risks of the hurdles becoming insecure—in fact, they will sometimes fall almost
of their own accord, on account of the loose state of the
soil. This occurs also at the break-up of a frost. Special
care is, therefore, needed on these occasions.

In winter, at the approach of frost, when severe weather
may be expected during the night, the stakes should be
loosened, or in all probability they will be frozen into the
ground ; and if the frost continues the hurdles will be
rendered useless until after the frost breaks up. In all
long frosts large numbers of hurdles are frozen in this
way, and the penning of the sheep becomes a matter of
difficulty.

Fold-Setting.

The shepherd has less hurdle-setting to perform if he
takes two pens instead of a single pen down the field at
once, as there are then only three sides instead of four

sides to set, the middle row forming an outside to both
pens. In Fig. 23 the usual plan of folding sheep is shown.

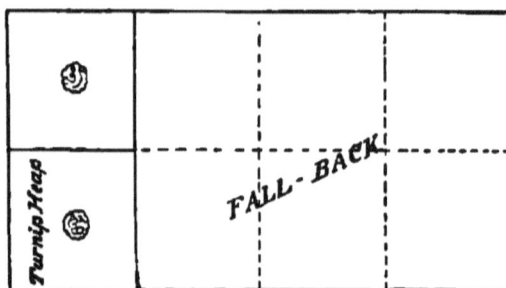

Fig. 23.—SHEEP FOLD, SHOWING FALL-BACK.

The full lines show the outsides of the folds; the dotted
lines indicate where the dividing hurdles have been, but
are removed to form a fall-back.

When setting hurdles the stakes should be placed on
the outer side, as greater resistance is obtained so, and
the shackles should be held tightly by the stake, as when

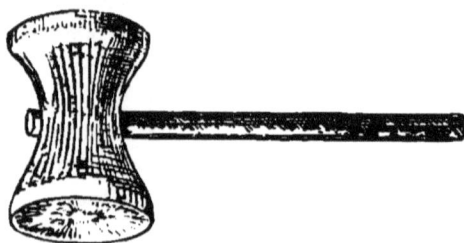

Fig. 24.—CUPPED IRON STAKE BEETLE.

loose the stake stands alone, but when tight it forms one
of a continuous range, each one giving support to the next.

When using the ordinary slatted hurdle with a stout

iron shackle to hold the fold stake, it is most convenient
to drive the stakes into the ground by means of a heavy
iron beetle, made cup-shape at each end (Fig. 24) so as to
prevent the stake from splitting by constant hammering.
However, in exceptionally dry or frosty weather a special
stake, shod with iron, called a drift stake (Fig. 25) must

Fig. 25.—Iron-shod Drift Stake.

be driven into the ground by means of a beetle to make a
hole in which to place the fold stake. The heads of the
hurdles should be driven into the ground as additional
support. Where wattled hurdles are used it is a common
practice to use a withy (twisted hazel or willow), or a
stout hempen shackle, which is slipped over the top of the
stake and twisted round the end upright spars of both
hurdles, and so made tight. In this case the hole is made
by an iron crowbar or fold bar.

In a few localities very heavy oak hurdles almost as
strong as gates are used, the heads being made exceptionally long, to admit of their being driven into the ground

Fig. 26.—Iron Fold Bar.

so as not to require the support of extra stakes, and an
iron fold bar (Fig. 26) being used to let the heads into the
ground. Although the hurdles are substantial and last for
a long time, they are not so profitable as the lighter forms,

as they are expensive to move. There is no doubt that
where used the hurdles are not moved as frequently as is
desirable, and they are not met with in the strictly close
folding districts, being more suited to use where little
moving is required.

Inequalities in the ground in a sheep fold are dangerous,
as sheep are liable to become "*cast.*" A sheep is cast
when it gets on its back and cannot get on its feet again.
Unless the wool is long and the sheep fat this rarely
happens on level ground ; but sheep with long wool,
particularly in warm weather in spring, when the ticks
become irritable and the sheep roll to get relief, very
frequently get cast, and, if not set on their feet, choke.
When the wind blows into their mouths, they choke
very quickly. On grass-land it is necessary to be among
long-wool sheep constantly in the spring or heavy loss
may be sustained. In-lamb ewes are liable to displace
the fœtus and lamb prematurely if cast. The danger of
casting is one reason why sheep lying in a big field
should be put up by the sheep dog when the shepherd
goes to them, as it is possible for him in a hurried
glance to overlook one, which would die if not turned
on to its feet.

The shepherd should see that the turnips are well
cleaned before being placed in the turnip cutter for
slicing. A piece of sickle or fagging hook makes the
best cleaner. When a shepherd receives no help he must
set the turnip cutter near to the heap of clean roots and
fill the machine by means of a spike driven into a short
stick, held in one hand, while he turns the machine with
the other. It is quicker and easier than filling the
hopper and then turning the machine. Kohl rabi are
very hard, and the root-stalk must be cut off : a skilful

man cuts these off as he chops the root with the long-handled adze usually employed for getting up the crop (Fig. 27), but a careless man leaves on a stump which is so tough as to make the slicing hard work.

Fig. 27.—Radi Chopper.

So much of the profit of sheep-keeping is derived from the *manure* that the management of the sheep must always be carried on with a view to distributing the manure so as to do the greatest good and the least harm. When feeding on pasture the troughs should be moved daily and not be kept near the gate to save trouble to a lazy shepherd. The sheep follow the troughs, and the manure is thus more equally distributed. The same applies to all cases where sheep lie out in large fields. Special pains must be taken on arable land where it is intended to take a corn crop subsequently. The troughs require constant moving, but another matter requires attention. When turnips are thrown into a heap, and the rootlets and dirt require cutting off before they are put into the turnip cutter, a heap of rich mould accumulates, and this, if not spread carefully about the ground, will cause a gross growth on the spot, probably resulting in the crop falling before harvest and becoming stained. These small patches of fallen grain are very difficult to keep separate, yet in the case of barley they will cause such an uneven sample that the corn will not realise so much by several shillings per quarter throughout the

field. Dirt and other accumulations in the troughs should
also be spread about and not allowed to lie in a heap.
For the same reason when sheep are placed in a small
pen for handling they should not be allowed to remain
too long, or the patch will become over-manured.

Unhealthy Foods.

A shepherd of experience generally knows to what he
should attribute any ordinary ailment. The matter of
paralysis through giving the sheep an over-supply of
nitrogenous matter has been gone into in a previous
chapter, and need not be enlarged upon here. If the
bowels are too much *relaxed* it is either due to too much
corn, too many roots in an unripe, rotten or other
unhealthy condition, or to something injurious in the
hay or chaff, though an internal chill may cause it. The
farmer and shepherd should at once look into the matter,
determine the cause, and rectify the diet. If the roots
are rotten or unripe, while the other food is good, the
cause may be expected there, and some portion should be
withdrawn, and the deficiency made good by bulky dry
food, such as hay or chaff. If the roots are good it
generally points to something wrong with the corn or
cake, which may have been heated at some time, or
become mouldy. This is frequently the cause, and when
it is requires correction. Mouldy hay may cause it. We
had a strange personal experience of poisoning on one
occasion when the whole of the ewe flock was affected.
The ewes were suckling lambs, which were fortunately
old enough to eat food in addition. The sheep suddenly
lost the use of their hindquarters, the milk dried up, and
they scoured badly, the fleeces being covered with coppery
scum and no little mucus. There appeared every likeli-

hood of their dying. The shepherd finding them in this state gave them more chaff, which, fortunately, they were unable to eat. It appeared that the shepherd, when going to the hay-stack, passed by where a stack of wheat was being threshed, and noticing what he thought was an agreeable smell from the chaff blown out of the machine, thought it would be an excellent food for the sheep. The sheep ate it freely, with the result mentioned. The sheep were given castor-oil, and then fed on rich linseed cake, split peas, and sweet hay, and recovered, not one being lost. The lambs were shut away from their mothers, and suffered very little, the drying up of the milk probably saving them. The milk gradually came back. The injurious matter in the chaff consisted of the seed-heads of stinking chamomile (mayweed) and wild marigold.

Mayweed often causes *sore noses* to sheep feeding on stubbles. The stubble pricks the noses of the sheep, and poison gets into them, causing festering sores and (sometimes) gums. Such sheep are best isolated and fed on easily eaten foods. A contagious form of sore nose is much more to be dreaded, especially where sheep gnaw roots, as the contagion is caught as the sheep go from root to root. Sore noses and mouths prevent the animals from getting a proper amount of food, and they lose condition, the effect on a whole flock being a very serious loss. Sore noses, from whatever cause originating, should be regarded with suspicion, and sheep affected should be withdrawn from the flock and isolated at once. Bad cases should be bathed, and cooling ointment applied.

The farmer should handle a few of the sheep at least once a week to see whether they are in an improving or a retrogressive condition, in order that he may regulate the food accordingly. It is generally sufficient to run them into a small pen, and handle them on the loin and dock. A man accustomed to handling sheep soon finds out whether the sheep are doing well or the reverse, and young farmers should practise this frequently to acquire the knowledge. The condition of the sheep indicates whether it is advisable to hasten them on to the butcher. Local trade for a special class of meat often regulates the time at which sheep should be got out, though, of course, the general market has to be regarded also. The amount of keep the farmer has at command must also be considered. If there is a likelihood of its running short, the obvious course is to feed the sheep in such a way as will fatten them while it holds out.

Selling Out Tegs.

Fat tegs are most generally sold off in small quantities as they become fit for killing. Store tegs are oftener sold in large numbers. It is frequently found advisable to divide the tegs in two or more lots, so that they may be hastened on as appears desirable. This may cause a little extra trouble to the shepherd, but it is well warranted. When sending fat sheep to market it is usual to sell them in small groups. They undoubtedly sell better when each pen is made up level by selection. Those of the same size, colour, type of head, and condition are best drawn together, as they look better and sell better so. Even when selling larger numbers they should be made to match. It is wise to draft out those which differ considerably in size, quality and type, and sell these separately. It may be taken as a safe rule that inferior sheep spoil the appearance of the better, and that the better do not improve the appearance of the poorer to a corresponding degree. If the sheep are sold in the wool, trimming off loose wool about the face, dirty locks of wool about the body, and a squaring of the tail always make them more "matchy" and more saleable; a rough un-groomed horse never sells as well as one smartly turned out, and this holds equally good with sheep.

Fat tegs which are *shorn* immediately before being sold should be done neatly and skilfully. Where the work is done carelessly, particularly if the back is not neatly finished, giving it a ridged instead of a flat appearance, the animal is placed in a prejudiced position, as it looks thinner and narrower than it really is. The legs and head also should be neatly trimmed.

An extra shilling paid for shearing a score of sheep well is money judiciously spent, as each sheep may fetch from a shilling to two shillings more. If branded with a pitch mark before going to market, a small neat brand should be used. A large brand carelessly put on detracts from the appearance. If fat tegs are shorn in cold weather, they should be kept warm, otherwise the meat becomes chilled and will not set properly. A supply of thick jackets made on the principle of a horse rug should be kept for the purpose.

CHAPTER XIII.

SHOW SHEEP.

Management of a Show Flock—House-Feeding—Trimming for Shows—Colouring for Exhibition.

Management of a Show Flock.

THE general principles in the management of a show flock do not differ greatly from those adopted in the case of any other well-managed flock, but greater care is exercised in details. Show flocks are as a rule essentially breeding flocks, as the expenses connected with exhibiting cannot be met unless other portions of the flock are sold at an enhanced price on account of the reputation made by those which have incurred extra expense in their preparation for exhibition. Exhibiting is, in fact, the most approved method of advertising the merits of a flock. It is therefore largely in the matter of getting up for exhibition that the difference between an ordinary well-managed flock and a show flock exists. This is not entirely the case, however, as the most valuable portion of the flock for sale and exhibition purposes consists of the rams.

It is useless, of course, to attempt breeding for exhibition unless good stock ewes and rams are kept. A change of blood has to be introduced from time to time to maintain vigour and improve the type; this is most economically done through the rams, as, though high-

HAMPSHIRE DOWN RAM LAMBS.

priced individually, they influence a large section of the flock. The occasional purchase of a few ewes from a specially good flock is desirable, as they are useful for mating with the rams of the home flock. The selection of the ewes is an important matter, as they are the permanent section of the flock, and are representative of its type and capabilities. The breeder's watchword must be "Improve," and he must aim at a type, and breed and select to that type. The rams must be purchased with that view, and must be selected to supply deficiencies and alter the character of features in accordance with the standard determined upon.

The soil has a great influence on the type, some kinds of land tending to the production of coarseness, others to too much fineness; and the variation is often observable on farms lying very near to one another. One farm is more favourable to the development of specially good rams, while another produces ewes of great merit. Where the ewes run fine a stronger type of ram is generally necessary, and where coarse a finer-bred ram. Where ewes are finely bred in-breeding is not so advisable as where there is a tendency to coarseness. In-breeding to a slight extent is generally productive of smaller and more finely bred animals. Carried beyond a certain point, the sheep become so fine as to indicate weakness.

The ewe sections of show flocks are usually rather better fed than are those kept for ordinary purposes, as the aim is generally to produce sheep of good size, although not at the expense of quality. Smallness does not necessarily indicate fineness of quality, and sheep may be coarsely bred though small. Small joints are cut from small sheep, but they may lack the quality of those from larger sheep. The ewe lambs and theaves, until they come into the flock, are usually kept in a fairly fresh condition. It is, however, the ram section that receives the most liberal treatment, and this is most marked in the case of the Hampshire sheep, of which a large proportion are used when lambs. In such instances the ram lambs have the first picking of all the best food, and all other sections of the flock are of secondary consideration, as they are put with the ewes when they are only eight months old. The ram lambs are fed with as much corn as they will eat from the time when they first feed until they are sold.

In those flocks where the rams are not used until they are shearlings it is not usual to feed them at such high pressure. A liberal allowance of corn is given while they are lambs, and this is continued so as to keep them in a thriving and growing condition until a few months before the sale, when they are fed on as heavy an allowance of corn as is consistent with safety, the object being to get them thoroughly fattened. It is found necessary by breeders to fatten their sheep more than is consistent with the activity and vigour looked for in a sheep going to service, because buyers will not recognise the valuable points of the animals unless they are developed. High feeding tends to the production of better wool, the appearance of finer quality of meat, a better outline, and more level handling, and shows definitely how much flesh the animal can carry. If these points are not developed buyers will not give credit for them, and the fault of over-feeding is one which is rendered necessary by the purchaser, and should not be ascribed to the feeder who does it on compulsion. It is, however, probable that the sheep are oftener unfertile on account of too much inbreeding than because of the over-feeding to which it is ascribed.

House-Feeding.

Although at one time it was strongly urged that the most profitable management of sheep was associated with house-feeding, the practice has not extended very largely, nor is it likely to. The cost of bringing bulky food to the sheep precludes it, except in special circumstances, such as the preparation of sheep for exhibition, or the production of "house-lamb."

House-feeding tends to the improvement of the quality of the wool, and the warmer atmosphere doubtless has an effect in making the animals fatten more quickly. The wool is also kept cleaner than when the sheep are fed in muddy folds; consequently a large proportion of the sheep exhibited are kept housed. The sheep-house should be well ventilated, and the floor kept dry and well littered with dry straw, otherwise foot-rot will break out.

House lamb is not so commonly produced as it was, as it is found the lambs thrive sufficiently well under ordinary circumstances. House lamb has generally been restricted to those which are born in autumn, and are required to be killed at Christmas or soon after.

Trimming for Shows.

The preparation for showing is an important matter. It is usual in the case of short-wool sheep to cut the wool on the back, so as to leave a square level surface, and also to cut in on other parts to give a good outline. Long-wool sheep are less clipped into shape, the object in their case being less to show the frame than the length and quality of the wool. In preparing the short-wool

K

sheep it is necessary to begin to level the wool several weeks before exhibiting, and to go over the work again at intervals.   Under the regulations of most Agricultural Societies sheep above one year old must have been shorn within some fixed period, generally the 1st of April, previously to being exhibited, otherwise the wool would be left on from year to year, and would not be truly indicative of the produce of the sheep under the conditions in which they are generally kept; there would also be a want of uniformity in appearance which would be misleading to those not well acquainted with sheep. In the eye of a good judge old wool is distinctly detrimental to the appearance of the sheep, as it lacks the bright lustre and freshness of newly-grown wool, and a man of even little experience can at once detect whether the sheep was shorn bare at the assigned time.

The sheep to be trimmed is placed in a rack specially made to hold it by the neck (in the clutches shown on the rack in Fig. 28), in a convenient position for the operator.

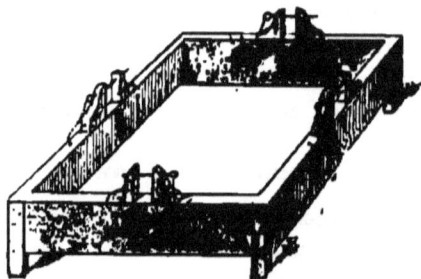

Fig. 28.—TRIMMING RACK, TO HOLD FOUR SHEEP.

It is first sprinkled with water, and the wool is cut off the back so as to give it a level appearance.   The wool is scraped with a wool sorter's comb, in order to break the locks and bring up the longer hairs, so that they may be trimmed off. The brisket is trimmed so as to give a bold, deep front. From the brisket to the neck a full outline is aimed at, the endeavour being to make an even curve, showing an unbroken sweep from the jaws to the brisket, in the same way that from the top of the shoulders to the poll the upper curve is made to show a natural sweep with no inequalities. The hind quarters are cut square, and to effect this it is generally necessary to leave a greater quantity of wool about the thighs than about the hips. The tail must be cut square; and though a full broad dock should be aimed at, it should not be allowed to hang so far back that it appears like an eave overhanging the rest of the hind quarters. The loose locks about the head and legs require clipping, and in most breeds the loose bunches are pulled out at the first trimming.

Trimming in this manner is performed on three or four occasions, so that the wool forms a dense pile. An experienced judge can tell how much the sheep is trimmed into shape by the varying appearance of the wool, as the lower it is cut on any particular part, the finer and closer it appears. Good trimming, like good grooming, adds to the appearance of the animal, but its chief value rests in the favourable impression it makes at first sight, thereby attracting attention.

Colouring for Exhibition.

It is the custom among exhibitors of several breeds to artificially colour the outside of the wool. Red or yellow ochre, mixed in oil, is commonly used, according to the fancy of the owner. Other breeds are merely oiled. Colouring with ochre is not a practice of recent introduction, as it was commonly practised in Hampshire more than a century ago. Its object is to give uniformity of appearance to a number of sheep together. The wool is so much influenced by the colour of the soil on which they have been folded, that when several lots from different coloured soils are brought together, some of them will appear of a more favourable colour, and will consequently give an impression of greater value than others, although they are in reality no better. Colouring the wool is, therefore, a justifiable practice, so long as it is not done to mislead. Where, however, colouring is applied to hide faults in breeding, it become reprehensible. For instance, dyeing the hair on the head, ears, or legs of sheep, which are not true in colour and marking, in order to make them appear so, cannot be too strongly censured, as also should be the eradication of rudimentary horns, because these practices are performed to hide blemishes in breeding, and to give the sheep credit for typical characteristics which they do not possess. The leading societies rightly make offences of this kind the ground for disqualification of exhibits.

The ochre is applied in the form of powder, or as a thin paste. When the colour is required to last more permanently, it is made into paste with oil, a small quantity being smeared on to the wool, and then rubbed on by hand. When the colour is not required to stand so long, a time when the wool is wet from light rain or heavy dew should be chosen; the powdered ochre should be sprinkled evenly over the sheep, through a flour-dredger, and then be worked in by the hand.

Sheep travelling to a show should be fed to a great extent on dry food during the previous day, to prevent scouring. It is advisable also to protect them by means of sheets, so that the wool does not become dirty or broken.

OLD NORFOLK EWE (3 YEARS), HER LAMB BY SOUTHDOWN RAM.

The property of Mr. Brown, of Norton, descended from the flock of Mr. Turner, of Creak.

CHAPTER XIV.

COMMON AILMENTS AND DISEASES.

Sturdy, or Gid—Liver Fluke, or Rot—Husk, or Hoose—Joint Evil, or Joint Ill—Scab—Abortion—Sheep Bot—Parasitic Worms—Louping Ill—Foot-Rot—Dew Rot—Foot-and-Mouth Disease.

Sturdy, or Gid.

SEVERAL ailments have been described in previous pages, but there are a few special ones which should receive more than passing mention. One of the commonest is known as sturdy, or gid, and arises from the presence of a parasite on the brain, which causes the sheep to assume peculiar actions. The early symptoms are restlessness and occasional shaking of the head. In a little time there is no mistaking the nature of the ailment; the sheep carries its head turned rigidly high on one side, and walks sturdily, but cannot pursue a straight course. Sometimes, but rarely, the head is protruded forward. As the disease progresses the animal experiences difficulty in feeding, as it cannot get its head down to eat, especially as at every attempt to move it turns round as though giddy. Later the sheep becomes dull about the eye and stupid. At this period symptoms not unlike those apparent when a sheep is partly paralysed from over-nitrogenous food are noticeable, it becomes listless, and loses power of locomotion. It then becomes a case for speedy cure or rapid wasting.   If the sheep is in good condition it may be killed; if not, the cause of the affliction should be removed.

The gid is caused by a parasite which has become encysted in the brain, and its seat can be detected by feeling, when an unnatural protuberance will be found. Under this is a cystoid or bladder-worm called *Cœnurus cerebralis*, which exists in a thin, transparent bladder-like form.   Sometimes more than one of these is found in one sheep.   The only remedy is to remove them.   This is done by trephining: an incision must be made in the skin and the skull, and the whole cyst carefully taken out.   The bone and skin must then be replaced, and kept bound up. As a rule cure and healing are effected.   Of course, the sheep must be isolated.   Young sheep and lambs are oftenest effected.

Prevention, however, is more satisfactory than cure. The bladder worm is one stage of the existence of a tape-worm called *Tænia cœnurus*, which inhabits the intestines of dogs.   A dog eating the brain of a sheep affected by gid cannot avoid eating the bladder worm, which finds a suitable resting-place in his intestines, although probably in only a small percentage of instances.   The worms produce eggs, and these fall on to pastures or other food of sheep, and are taken in with the food.   It has been suggested that they reach the brain through the nose, but there appears to be doubt as to the exact means by which they get there; that the tape-worm in the dog and the bladder worm in the sheep's head are different stages in the life of the parasite, however, there can be no doubt. Whenever a sheep dies from gid its head should be burned.

Sheep dogs should be systematically treated for worms,

and when under treatment should be kept confined; all worms and droppings coming from them should be burned. This applies to all dogs about a farm. Sheep fed on pastures where dung from kennels has been placed stand a great chance of becoming attacked.

Liver Fluke, or Rot.

No disease entails greater loss among sheep in all parts of the world, though more particularly in the moister climates, than the rot. It is caused by a small trematode worm, which is found chiefly in the liver. Until within comparatively recent years the life-history of this parasite was not known, and owing to the fact that it passes through several transformations, part of them not in association with the sheep, it was difficult to trace them. The fluke parasite (*Fasciola hepatica* or *Distoma hepatica*) does not enter the sheep in the form in which it is usually found. The fluke, whilst in the sheep, produces many eggs, which pass out with the droppings; these subsequently hatch, and the embryo attaches itself to the body of a snail, in almost every case to a special variety, *Limneus truncatulus*, which possesses a pale buff spiral shell not more than a quarter of an inch long. Further changes are undergone whilst it is a parasite of the snail, and the sheep in gathering their food take in the snail or some of the detached parasites, which develop into the liver fluke, and the functions of the liver are destroyed.

There is no known cure. Prevention is the farmer's only method of dealing with it. Low-lying grass land frequently under water is almost always an infected area, the exception being salt marshes, which, as salt is prejudicial to the snail, is safe except from an overflow of fresh water. The snail is found in almost every ditch, and when these overflow the snails are deposited on the flooded ground.   Sheep grazing along ditch sides often contract the disease even in dry seasons.   The greatest risk is run in autumn, and low-lying grass land should not be fed by sheep from July until after a severe frost.   In very wet years the snails creep along the moist grass to a considerable distance, and in the wet year 1879 they were found on turnips on comparatively high ground.   Salt is prejudicial to their development, consequently a supply should be before the sheep at all times.   Lime is prejudicial, but its practical application is difficult, and cannot be regarded as of great value.   The great point to attend to is to keep the sheep well supplied with food and healthy in other respects.   For a short time after attack the sheep fatten rapidly, and they should then be sold, for if left too long they will suddenly lose flesh, and in a short time waste away and die.   When the disease has obtained a strong hold no treatment avails.   We lost over 200 ewes from this disease in the winter 1879–80. In the hope of getting them to produce their lambs everything was done to maintain strength, but it proved useless, as all died either before or during lambing.

The disease generally becomes apparent from November to January, when the sheep become lazy, and for a short time fatten quickly.   The mucous membrane about the eyes, mouth, and nose loses its pinky colour and becomes pale.   This is the time to kill them, as in the course of a few weeks they will waste away and die.   If allowed to advance far, the wool may be plucked out as readily as down from a goose, and the skin is of little value.

In wet seasons the farmer should be very careful that when purchasing he avoids sheep which have been fed on rotting ground.

Husk, or Hoose.

In the autumn lambs are often afflicted with a husky cough, and if the windpipe is examined colonies of small worms are found in it. Considerable loss is experienced when the lambs are in poor condition, but if vigorous and well fed they generally succeed in coughing them out. The worm (*Strongylus filaria*) is a thread worm from three-quarters of an inch to two inches in length. It is generally found on wet pastures, but there it is in another form. The husk worm, like the liver fluke, does not spend its entire life in the sheep, but a portion of its career is spent in the common earthworm. The ova of the husk worm are deposited on pasture, and enter the earth worm, where embryo worms are developed. These pass out and attach themselves to moist grass, on which they are able to travel by an eel-like motion. From this they get on to the noses of the sheep feeding the grass, and pass up them, and so into the air passages : there they rapidly reproduce young, which are developed from time to time.

The dews of autumn afford a special opportunity for the worms to move, whereas in dry weather they cannot climb up the stems of the grasses ; consequently autumn, with its fogs and dews, is the most dangerous period. Land once infected generally remains so until it is well drained, and sheep should be kept off it for some time if attempts are made to free it from infection. Salt is prejudicial to the worms, and this is another reason for keeping salt in the sheepfold at all times. Rich food is the best means of fortifying the sheep against the attack.

Turpentine has a good effect in destroying the worms. A drachm of turpentine, with an egg beaten up in milk, is a good remedy, easily administered by means of a small drinking horn or strong-necked bottle. Inhalation of the vapour of tar, carbolic acid, and chlorine is effectual; this must be done in an enclosed chamber. When using chlorine care must be taken that the animals are not left too long, as they become suffocated through the collapse of the larynx. Bleaching powder, with hydrochloric acid poured over it, is the usual method of providing the chlorine fumes. An attendant should enter the chamber with them, and as soon as he feels distressed by the fumes he should open the door and allow the sheep to escape.

During the last few years a hypodermic syringe, with a strong, sharp needle-point, has been used for injecting turpentine into the windpipe. This is a very effective method, but no small risk is run, as the sharp prick makes the animal start, and often causes the rings of the trachea to be displaced, which sets up inflammation and may result in death. At any rate, the animal is much distressed. As the worms develop at intervals one application of any kind of preventive is not sufficient. Two or three, with about three days' interval between, are as a rule necessary.

Joint Evil, or Joint Ill.

One of the greatest troubles in the lambing yard, and for some little time after, is a swelling of the leg joints, often attributed to rheumatism; however, as it appears in the form of an epidemic, this is to an extent unreasonable, although as so many outward symptoms are suggestive of rheumatism it is perhaps excusable. It is now generally accepted that it is caused by a virulent organism capable of reproducing itself in the body, and that it finds its way into the system through an imperfectly closed navel. Blood-poisoning is set up, and the tissues about the joints are invaded, resulting in swellings, which are the outward indications of the disease raging within.

As loss from this disease is so great, preventive means should be adopted. Cleanliness is the first point to attend to, as a foul lambing pen is the source of the disease. The navel should be washed with an antiseptic solution (carbolic acid and olive oil in the proportion of one to seven), and induced to heal, so as to destroy the possibility of the germ establishing itself there. It is quite possible for the disease to be transmitted at a time of castrating or tail cutting, and this is a feature in favour of searing. The application of an antiseptic dressing as above acts both as a preventive and as a remedy, and is a better course to adopt. Strong lambs are much less likely to develop the disease than are weakly ones, and those chilled by wet or cold are especially liable. Generous feeding and the provision of warm shelter are therefore necessities. Foul pens and stale litter are the shepherd's greatest enemies.

Scab.

Owing to vigorous laws relating to infectious diseases this scourge of the sheep farmer is far less prevalent than in the past. Dipping in the recognised sheep-dipping solutions is responsible for its decrease and for the ease with which it is kept in check. It is a most troublesome disease when once it appears on the farm, and all new purchases should be dipped, if for the sole reason of preventing its introduction. The old practice of dressing with mercurial ointment, which was both troublesome and dangerous, is quite unnecessary. The disease is a form of mange mainly caused by a minute acarus (*psoroptes*, var. *ovina*), or mange mite, which burrows into the skin, sets up irritation, and develops scabby eruptions. A sheep affected shows signs of it by gnawing

at the wool, and by scratching itself. If the fingers are
inserted into the wool and the skin is rubbed, the scabs
are easily felt; the sheep turns its head in the direction
of the sores, and moves its mouth in a quick nibbling
manner. The disease is very contagious, and the shepherd
should always regard a broken-coated sheep with suspicion.

Isolation of the affected sheep should be at once effected,
and the flock should be dipped without delay. Three or
more dippings may be required to effectually clear the
sheep, as the smallest patch left untouched is sufficient
to set up disease again in the near future. Where the
disease has broken out all loose locks of wool, whether on
hurdles, hedges, or elsewhere, should be collected and
burned, as the disease is carried in them.

A convenient dipping apparatus is shown in Fig. 29.
The sheep, held by its legs by two men, and one man holding

Fig. 29.—SHEEP DIPPING APPARATUS.

the head to keep it from swallowing the solution, is put
in the bath back downwards for a minute. It is then laid
on the draining rack while the liquid is squeezed out.
Cooper, McDougall, and several other well known firms
supply good "dips."

Abortion.

A number of sheep produce their lambs prematurely. Many of these cases result from injudicious feeding, and from accident, but by far the greater number are traceable to contagion, when the affection takes an epizootic form. The effect of injudicious feeding has been described previously. Abortions occasioned by accident generally arise from the sheep being strained through over-exertion, as when made to travel too fast, or when hunted by dogs. Probably the effort of walking through pens deep in sticky mud entails an undue strain on them, which puts the lamb out of its proper position. Rushing sheep through narrow gateways is a not uncommon cause when the young lamb has acquired much of its growth. Turning the sheep on to its back is another frequent cause. Sympathy is often ascribed as a predisposing element, though probably since the fact has been established that it is largely brought about by a specific germ, this cause is less often accepted. Ergotised grasses, largely blamed a few years ago, is little entertained now as a cause. It is now conclusively shown that when a flock is seriously affected by abortion it is more often caused by a specific germ, which is highly contagious.

In a case of abortion all parts expelled should be burned, as should any litter or other material which is brought into contact with it. The ewe should be kept apart from others until all traces of the matter have left her, and the hindquarters should be well washed and bathed with a solution composed of ten pints of rain water, 2½ drachms of corrosive sublimate, and 1½ oz. of hydrochloric acid. It is perhaps more important to give attention during the early portion of the period of gestation, as there is greater risk of others being affected. An ewe which has aborted should be fattened, for she will probably abort again, as the contagion often remains in her.

Sheep Bot.

The sheep bot is the larva of the sheep bot fly (*Œstrus ovis*), which lays its eggs in the nostrils of the sheep during hot weather in July and August. When the fly is active the sheep show signs of terror, and huddle together for protection. They put their heads between their legs, and keep their noses near the ground, to prevent the fly from entering the nostrils. Serious injury is not often occasioned by the bot, though sufficient inconvenience is effected to cause the sheep to be somewhat restless.

McDougall's Carbolic Smear or paraffin rubbed on the noses of the sheep every few days when the attack is probable, prevents the fly from laying its eggs in the nostril. Tar is used for the same purpose. An ingenious method is to place a block of rock salt behind a board through which augur holes two inches in diameter are bored, and to keep the hole smeared with tar. The tar gets on to the nose every time the sheep licks.

Parasitic Worms.

The mystery which has so long prevailed as to why stale keep affects sheep injuriously, appears to be nearer solution than seemed likely a few years since. The great advance made in veterinary science has led to the sheep coming more closely under scientific observation. Veterinary surgeons had not devoted much special attention to the sheep, because they were so rarely called in to prescribe, and they felt the time devoted to the investigation of the animal to be of little pecuniary value. Recent investigations, however, indicate that they will become much more indispensable to the sheep farmer in the future; and such being the case, there is no doubt that with scientific aids they will throw light on matters which have hitherto baffled the shepherd. It is true that some of their discoveries have been so recent that, although in some cases the causes have been ascertained, adequate remedies have not been formulated at present: but it can scarcely be doubted that before long remedies will be available.

Many of the afflictions of sheep which have baffled the shepherd have now been shown to be due to various internal worms. One of the most pernicious of these is the small nematoid worm, found in the fourth stomach of the sheep, and called *Strongylus cervicornis*, which only under the most favourable circumstances can be detected by the naked eye, and which, until recently, had passed unnoticed, although frequently present in sheep in myriads. Its presence is associated with scouring, rapid wasting, loss of appetite, and craving for water; and the sheep frequently display an inclination to lick sand or earth. It most commonly affects lambs, and in the large number of cases in Lincolnshire investigated by the authorities at the Royal Veterinary College in 1896, deaths ranged from five to twenty per cent.

In addition to the small nematoid worm just mentioned, there are tapeworms infesting the digestive channel, some of which are more easy of expulsion.

Besides the *Strongylus filaria*, already discussed in connection with husk, which inhabits the tubes of the lungs, there is the *Strongylus rufescens*, a red thread worm, which is found in patches or nodules in different parts of the structure of the lungs. Also the hair lung worm (*Pseudalius ovis*), which is coiled up in small tubercle-like spots throughout the lung-structure, and is often very abundant on the surface directly underneath the pleural membrane. The presence of these worms has been recognised for some years, but they have been to some extent confounded until recently.

It is not uncommon for land to become what is known as "sheep sick," because sheep fail to thrive on it when they are kept on it too thickly and too frequently : but why this should be was not understood, many theories being suggested, but these not of such a nature as to stand thorough investigation. On farms mainly given up to sheep, farmers have found the advantage of breaking up the land to grow cereal crops to "sweeten the soil," rather than keeping it always under root and forage crops for consumption by sheep. This practice will have to continue. How far the burying of the worms or their embryos helps to destroy them, is not proved, though, doubtless, the point will be investigated. Some of these worms do not reproduce mature worms inside the sheep, and the number of worms inhabiting a sheep must always be dependent on the number taken in, even though a portion of them may rapidly be expelled. Therefore the land must not be overstocked. Prevention will have to be in the direction of rendering the worms scarce in the fields. As they appear to thrive in moisture, one of the first objects must be to drain the land. The cropping must be diversified, so that sheep are not on it too frequently. It is probably because of the dry nature of the soil, and the frequent ploughings for catch crops, that the Down land carries so many sheep healthily; though Down land is not by any means free from the pests. Exposure and shortness of keep tend to the debility of the sheep, lessening its chances of resisting attacks.

On pasture farms, it is obvious that as the land must be left down in grass permanently, it is desirable to rest the infected fields for a time, substituting bullocks for sheep where the liability to attack has become so serious that sheep keeping cannot be carried on except at the risk of great loss.

Louping Ill.

Louping ill is a most serious disease, rarely found in England, but not by any means uncommon in Scotland, and occasionally met with in Ireland. The affected sheep travels with difficulty, being affected with partial paralysis, most frequently in the hind quarters. The animal occasionally leaps forward, and then is overcome with twitching and trembling. The eyes become glassy and the ears droop. The appetite becomes depraved, and the sheep eats dirt or anything that comes before it.

Healthy management, good food, as little disturbing or frightening as possible, and shelter in cold, wet weather at lambing time, are practically all that can be advised. Fat sheep should at once be killed.

A vast number of causes are ascribed as producing this disease, but it is premature to assert with confidence that it is definitely known what is actually and universally the cause. Many hold that it is a form of ergotism, caused by the sheep eating grass infested by ergot; but this is not satisfactorily shown; and there seems a likelihood that the tick theory is nearer the mark. The disease has certainly been found most prevalent in years when ticks have been unusually abundant. Microscopic organisms have been found in the fluid surrounding the

L

spinal cord of affected sheep, and similar organisms have been found in ticks on sheep suffering from the disease, so the natural inference is that they inoculate the sheep. Some support for this is found in the *tick fever*, which is doing so much harm among cattle in Queensland. In that case, however, sheep are almost the only animals which are not there troubled with the special kind of tick. While ticks are found infested with the organism, it is certainly advisable to get rid of them as effectually as possible, and dipping should be carefully and frequently done. The disease is such a serious hindrance to sheep-farming in Scotland, that it is desirable that no reasonable cost should be spared to discover everything possible as to its origin, prevention, and cure. England has kept remarkably free from it, and great care should be exercised against its introduction; considering the large and increasing number of sheep sent south, it is astonishing that it should not have become established among English flocks. If ticks are the source of the disease, no sheep should be allowed to cross the Border unless previously and recently carefully dipped.

Foot-Rot.

This subject has been referred to frequently, but there are a few points which could not be conveniently dealt with, and which may be treated here.

Foot-rot, as distinct from lameness caused by injury or neglect, is contagious, being caused by micro-organisms which apparently are able to exist in the soil for a lengthened period, and are very difficult to destroy. We know of definite instances of farms on which the disease had been unknown until it was imported, but which have since always retained the contagion. Experiments have proved the contagious nature of the disease. Wet weather and wet lair seem to· have a marked influence on the development of the disease, but whether it is due to the greater activity of the germs, or entirely to the softened and more easily abraded skin and hoof, is not definitely proved. The disease is set up on the skin between the claws, and, if not checked, it spreads down into the hoof, separating the hoof from the foot, or on to any part of the foot where the hoof has broken away.

Sheep lying on soft, wet, arable ground have little opportunity of getting hard feet, and should occasionally be driven on hard roads, or be placed for a short time on dry caustic lime; the latter has an excellent effect on the feet and the skin about the hoof. When foot-rot shows itself in a flock, prevention may be effected by dressing the skin about the coronet and between the claws with Stockholm tar. A quick and easy method of dressing the feet is to make the sheep walk through a trough, in which is a solution of arsenic, twice a year, with a view to hardening the feet, and where foot-rot breaks out to give them another bath in the same way to cure it. Previously to walking them through the solution, the feet should be pared into shape. The solution should be composed of 2lb. of arsenic and 2lb. of pearlash, boiled in a gallon of water over a slow fire for half-an-hour, after which, five gallons of water should be added. A stout trough, 12ft. long by 18in. wide and about 6in. high, should be provided, at the bottom of which should be placed cross-pieces to prevent the sheep from slipping; into this sufficient solution to just cover the hoofs of the sheep should be put. The trough requires placing in a division between two adjoining folds, and the folds should be made on fallow land, so that no food is poisoned by the solution.

The most convenient position to place the trough is in an opening made in the dividing fence close to one side, but a fence must be run alongside the trough parallel with the outer fence, so that when once the sheep is in the trough, it will have to walk through from end to end. To prevent rushing and injury to the sheep, a small division may be made, so that only one sheep reaches the trough at once, and it is best in the form of a narrow alley, made by another line of hurdles inside the pen, but opening wider to form an inlet (Fig. 30). No handling of

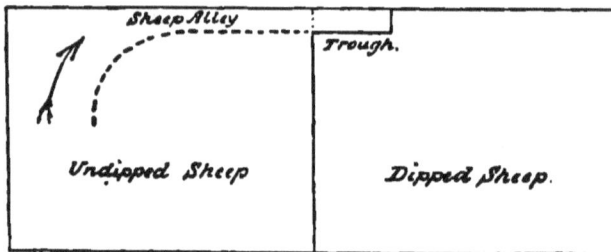

Sheep Alley

Trough.

Undipped Sheep   Dipped Sheep.

Fig. 80.—Fold adapted for Passing Sheep through Foot-Rot Bath.

the sheep is necessary: they are simply allowed to pass through the trough; though a swinging hurdle may be placed at the outlet to make them pass through steadily, otherwise they will splash out a lot of the solution without properly soaking their feet. As the solution is *poisonous*. the sheep should not be taken out of the fallow fold for half an hour or more, or the food would be poisoned; nor should the liquid be left exposed at any time.

The reason why so many shepherds fail to get rid of foot-rot within a reasonable time is that they do not pare the sheep's feet sufficiently. Unless the caustic portion of the remedy reaches all parts where there is disease, it is impossible for it to effect a cure. The least speck of disease is sufficient to re-start foot-rot, yet the hoof may appear perfectly sound for several weeks, as owing to the horn of the hoof having become hardened, it cannot break through in the ordinary manner, but bursts out at the top. This is always a bad case, as it diseases the upper portion, and a new hoof has to be grown. While following up the disease by paring the hoof where it is required, the brutal cutting which some shepherds inflict is in no way justifiable. There is no necessity to make the foot bleed, and it does no good. All that is necessary is to pare the hoof wherever it has become loosened from the foot. If this is done, the caustic can easily be applied to all diseased parts, and the disease be effectually cured. The ignorant idea that it is necessary to "sting up" the foot is most cruel and unwarranted. It is impossible to cut the hoof properly without a strong, sharp knife. Macklin, of Salisbury, makes one specially for the purpose, and it can be strongly recommended.

Dew-Rot.

Dew-rot is a disease affecting the hoof, causing it to fall off if left too long. Foot-rot often develops with or after it, as the raw skin is a convenient place for it to establish itself. In the case of dew-rot white blisters form about the top of the foot and between the claws, the hair falls off, and the blisters become open sores.

The sores should be opened, and stimulating embrocation applied. If the case proceeds so far that the hoof slips, the foot must be bathed, poulticed, and bound up to keep out dirt. The disease is commonest in spring and autumn, but it may break out in wet summers. It most frequently attacks sheep on wet pasture, and they should be put on dry arable land.

Foot-and-Mouth Disease.

Thanks to stringent regulations, this disease is likely to do less injury than in the past fifty years. Owing to its afflicting all farm stock, except horses, it may be regarded as one of the most serious diseases of the farm. It is less fatal than some diseases, but the "wasting" which it inflicts on the animals, and the length of time which elapses before they regain flesh, cause it to be dreaded more than almost any other. Foot-and-mouth disease (epizootic aphtha) is an eruptive fever produced by a specific germ, and is highly contagious. Although fever and a staring coat are premonitory symptoms, its presence is not usually noticed until attention is drawn to the sheep's mouth, from which ropy saliva flows, the animal champing its jaw and making a sucking noise. If the mouth is examined, blisters will be found in it, and the animal feeds with great difficulty. White blisters are formed about the top of the hoof where it joins the skin, and these soon break, leaving open sores. Sometimes the mouth is little affected, while the feet are badly ulcerated.

Salicylic acid in solution (1 to 300 of water) is an excellent mouth-wash, and the feet should be dressed with a solution of alum and carbolic acid. The animals should be kept warm, the lair should be dry, and soft, rich food in the form of scalded mashes should be given while the mouth is tender. Everything should be done to maintain the strength, though in cases of severe constipation a mild aperient is beneficial. As the foot is in a more or less ruptured condition, great care is necessary for some time to guard against foot-rot, which very frequently follows the disease.

DORSET HORNED EWES.

CHAPTER XV.

Ear-Marking—Sale and Purchase—Choice of a Sheep-Dog.

Ear-Marking.

IT is desirable to mark the sheep in a permanent manner, and wool marking by means of a pitch brand does not do this, although it is very useful as a temporary marking. The ear is the best place, and it lends itself to the purpose in several ways, such as tattooing, slitting, and punching. Tattoo punches prick through the skin, and leave an impression similar to the pattern of the punch, so that when coloured matter is rubbed in, it becomes permanently fixed. Numbers or any special design can thus be placed on the sheep.

Slits or nicks on the edge of the ear can be made to indicate anything, according to the code the farmer may adopt. A nick on a certain side of a particular ear may be made to indicate the ram by which the sheep was got, and the year it was got. As there are two ears with two edges, and as three or four nicks may be made on either side, there is sufficient scope to meet the necessities of a large flock.

One or more holes may be punched in the ears to identify sheep. In pedigree flocks buttons with studs to keep them in place are used: these are placed in holes previously punched through the ear, and as any number

151

or design can be stamped on them when being made, an easy method of keeping a record is obtained. The number should be recorded in the flock book. In the case of pedigree flocks, the buttons should be made to close in such a way that they cannot be removed or used on a second occasion. A strip of soft metal is often used, the ends of which are inserted through two slits in the ear, and then turned so as to prevent slipping out. Numbers can be stamped on these to identify the sheep.

Sale and Purchase.

In previous chapters remarks have been made on points to be looked for in handling a sheep, and on the necessity of making even drafts when offering them for sale. There remain, however, a few points which may well be dealt with. So many animals now pass through the auctioneers' hands, that there is little doubt farmers have lost some of the skill they possessed in estimating the weight of their sheep when they more frequently sold them to the butcher. Butchers realize that they can buy cheaper from the auctioneer than from shrewd owners, and naturally purchase from the former. On the other hand, the farmer is sure of prompt payment from the auctioneer, whereas the butcher was not always ready to pay on the deal. He, however, pays the auctioneer pretty heavily for the accommodation. It is a recognized axiom that animals always show themselves better on the farm than when tired and jaded in the market.

Selling is effected in several ways, by auction, upon legs, by live weight, or by dead weight. When sold *by auction* the farmer pays a percentage on the sale. "*Upon legs*" is a term used to denote they are bought as they stand, as when sold from the fold; a lump sum being given for the sheep as they are, with no deductions or additions. When sold by *live weight* they are disposed of at so much per stone, gross weight. When sold by *dead weight* they are sold per stone dead weight. It is usual to sell them *sinking the offal*: this implies that the wool, skin or pelt, head, and the whole of the internal organs with the exception of the kidneys and the kidney fat or suet, and the legs below the knees and hocks, are not weighed in, but a sum equivalent to their value is mentally calculated, and added to the price per stone of the whole carcass.

A considerable difference is, of course, made when sold carrying a heavy fleece or freshly shorn. A twelve pound fleece, at 10*d*. per lb., naturally adds ten shillings to the value of a sheep, and if the sheep weighs ten stones, it makes a difference of a shilling per stone. The dead weight of an animal is taken when the body has cooled and dried. As the sheep are killed from home, it is necessary for the seller or his trusty agent to see them weighed. A simple brand should be put on the heads, and the head should not be severed from the carcass until the time of weighing, as there is no other way of identifying the body, and another may be substituted to the disadvantage of the seller.

When buying by live weight, it is usual to calculate on the fasted live weight, as it is very different when calculated on an animal with its paunch full. A rough estimate is found on the basis of allowing 8lb. of carcass and 6lb. of offal to each stone of 14lb. It is usual, therefore, to speak of a *stone live weight* as being 14lb., and a *stone of mutton* as 8lb., the two being described as *long* or *live weight*, and *short* or *dead weight stones*. This is by no means applicable in all cases, as the percentage varies greatly between a store and an ordinary fat sheep, and of course, far more in the case of a sheep fit for exhibition purposes. Condition and breed are matters to be considered. Down breeds, which carry proportionately more lean meat, weigh heavier than long-wool white-faced breeds, but under ordinary circumstances the following calculations may be taken as fairly representing the proportion of mutton to live weight of sheep.

Live Weight in Pounds.	Percentage of Mutton.	
	In Wool.	Newly Shorn.
280 to 300	61 to 72	74 to 75
260 „ 280	69 „ 70	73 „ 74
240 „ 260	67 „ 68	71 „ 73
220 „ 240	65 „ 66	69 „ 70
200 „ 220	63 „ 64	67 „ 68
180 „ 200	61 „ 62	65 „ 66
160 „ 180	59 „ 60	64 „ 65
140 „ 160	58 „ 59	63 „ 64
120 „ 140	56 „ 57	62 „ 63
100 „ 120	55 „ 56	60 „ 61
80 „ 100	53 „ 54	58 „ 59
60 „ 80	50 „ 52	56 „ 57

Skins of sheep, whether with wool on or not, should be sold when fresh. It is a great mistake to leave them to spoil, as is so commonly done.

Choice of a Sheep-Dog.

In conclusion, the shepherd requires a good dog, be it collie, Old English, or mongrel. Without a dog shepherding is most troublesome, hardly less so where sheep are close folded than where they are running on mountain or heath. Time devoted to the careful training of a young dog is therefore well spent. It should be quick but gentle, ready to bark but slow to bite, obedient, but resourceful through its own intelligence when out of sight of its master. The easiest plan is to break it with an old dog to set the example under varying circumstances, and the earlier it is brought among sheep the better, for it is difficult to break an old one which has formed habits of its own.

INDEX.

Catalogue of Practical Handbooks

Published by L. Upcott Gill, 170, Strand, London, W.C.

CONTENTS.

American Dainties, and How to Prepare Them. By an AMERICAN LADY. *In paper, price* 1s., *by post* 1s. 2d.

Angler, Book of the All-Round. A Comprehensive Treatise on Angling in both Fresh and Salt Water. In Four Divisions as named below. By JOHN BICKERDYKE. With over 220 Engravings. *In cloth gilt, price* 5s. 6d., *by post* 5s. 10d.
    Angling for Coarse Fish. Bottom Fishing, according to the Methods in use on the Thames, Trent, Norfolk Broads, and elsewhere. New Edition, Revised and Enlarged. Illustrated. *In paper, price* 1s., *by post* 1s. 2d.
    Angling for Pike. The most approved Methods of Fishing for Pike or Jack. New Edition, revised and enlarged. Profusely Illustrated. *In paper, price* 1s., *by post* 1s. 2d.
    Angling for Game Fish. The Various Methods of Fishing for Salmon; Moorland, Chalk-stream, and Thames Trout; Grayling and Char. Well Illustrated. *In paper, price* 1s. 6d., *by post* 1s. 9d.
    Angling in Salt Water. Sea Fishing with Rod and Line, from the Shore, Piers, Jetties, Rocks, and from Boats; together with Some Account of Hand-Lining. Over 50 Engravings. *In paper, price* 1s., *by post* 1s. 2d.

Angler, The Modern. A Practical Handbook on all Kinds of Angling. By "OTTER." Well illustrated. New Edition. *In cloth gilt, price* 2s. 6d., *by post* 2s. 9d.

Aquaria, Book of. A Practical Guide to the Construction, Arrangement, and Management of Freshwater and Marine Aquaria; containing Full Information as to the Plants, Weeds, Fish, Mollusca, Insects, &c., How and Where to Obtain Them, and How to Keep Them in Health. By REV. GREGORY C. BATEMAN, A.K.C., and REGINALD A. R. BENNETT, B.A. Illustrated. *In cloth gilt, price* 5s. 6d., *by post* 5s. 10d.

Aquaria, Freshwater: Their Construction, Arrangement, Stocking, and Management. By REV. G. C. BATEMAN, A.K.C. Fully Illustrated. *In cloth gilt, price* 3s. 6d., *by post* 3s. 10d.

Aquaria, Marine: Their Construction, Arrangement, and Management. By R. A. R. BENNETT, B.A. Fully Illustrated. *In cloth gilt, price* 2s. 6d., *by post* 2s. 9d.

Australia, Shall I Try? A Guide to the Australian Colonies for the Emigrant Settler and Business Man. With two Illustrations. By GEORGE LACON JAMES. *In cloth gilt, price* 3s. 6d., *by post* 3s. 9d.

Autograph Collecting: A Practical Manual for Amateurs and Historical Students, containing ample information on the Selection and Arrangement of Autographs, the Detection of Forged Specimens, &c., &c., to which are added numerous Facsimiles for Study and Reference, and an extensive Valuation Table of Autographs worth Collecting. By HENRY T. SCOTT, M.D., L.R.C.P., &c. *In leatherette gilt, price* 7s. 6d. *nett, by post* 7s. 10d.

Bazaars and Fancy Fairs: Their Organization and Management. A Secretary's *Vade Mecum.* By JOHN MUIR. *In paper, price* 1s., *by post* 1s. 2d.

Bees and Bee-Keeping: Scientific and Practical. By F. R. CHESHIRE, F.L.S., F.R.M.S., Lecturer on Apiculture at South Kensington. *In two vols., cloth gilt, price* 16s., *by post* 16s. 6d.
    Vol. I., Scientific. A complete Treatise on the Anatomy and Physiology of the Hive Bee. *In cloth gilt, price* 7s. 6d., *by post* 7s. 10d.
    Vol. II., Practical Management of Bees. An Exhaustive Treatise on Advanced Bee Culture. *In cloth gilt, price* 8s. 6d., *by post* 8s. 11d.

Bee-Keeping, Book of. A very practical and Complete Manual on the Proper Management of Bees, especially written for Beginners and Amateurs who have but a few Hives. By W. B. WEBSTER, First-class Expert, B.B.K.A. Fully Illustrated. *In paper, price* 1s., *by post* 1s. 2d.; *cloth*, 1s. 6d., *by post* 1s. 8d.

Begonia Culture, for Amateurs and Professionals. Containing Full Directions for the Successful Cultivation of the Begonia, under Glass and in the Open Air. By B. C. RAVENSCROFT. New Edition, Revised and Enlarged. Illustrated. *In paper, price* 1s., *by post* 1s. 2d.

Bent Iron Work: A Practical Manual of Instruction for Amateurs in the Art and Craft of Making and Ornamenting Light Articles in imitation of the beautiful Mediæval and Italian Wrought Iron Work. By F. J. ERSKINE. Illustrated. *In paper, price* 1s., *by post* 1s. 2d.

Birds, British, for the Cages and Aviaries. A Handbook relating to all British Birds which may be kept in Confinement. Illustrated. By DR. W. T. GREENE. *In cloth gilt, price* 3s. 6d., *by post* 3s. 9d.

Boat Building and Sailing, Practical. Containing Full Instructions for Designing and Building Punts, Skiffs, Canoes, Sailing Boats, &c. Particulars of the most suitable Sailing Boats and Yachts for Amateurs, and Instructions for their Proper Handling. Fully Illustrated with Designs and Working Diagrams. By ADRIAN NEISON, C.E., DIXON KEMP, A.I.N.A., and G. CHRISTOPHER DAVIES. *In one vol., cloth gilt, price* 7s. 6d., *by post* 7s. 10d.

Boat Building for Amateurs, Practical. Containing Full Instructions for Designing and Building Punts, Skiffs, Canoes, Sailing Boats, &c. Fully Illustrated with Working Diagrams. By ADRIAN NEISON, C.E. Second Edition, Revised and Enlarged by DIXON KEMP, Author of "A Manual of Yacht and Boat Sailing," &c. *In cloth gilt, price 2s. 6d., by post 2s. 9d.*

Boat Sailing for Amateurs, Practical. Containing Particulars of the most Suitable Sailing Boats and Yachts for Amateurs, and Instructions for their Proper Handling, &c. Illustrated with numerous Diagrams. By G. CHRISTOPHER DAVIES. Second Edition, Revised and Enlarged, and with several New Plans of Yachts. *In cloth gilt, price 5s., by post 5s. 4d.*

Bookbinding for Amateurs : Being Descriptions of the various Tools and Appliances Required, and Minute Instructions for their Effective Use. By W. J. E. CRANE. Illustrated with 156 Engravings. *In cloth gilt, price 2s. 6d., by post 2s. 9d.*

Breeders' and Exhibitors' Record, for the Registration of Particulars concerning Pedigree Stock of every Description. By W. K. TAUNTON. In 3 Parts. Part I., The Pedigree Record. Part II., The Stud Record. Part III., The Show Record. *In cloth gilt, price each Part 2s. 6d., or the set 6s., by post 6s. 6d.*

British Dragonflies. Being an Exhaustive Treatise on our Native Odonata : Their Collection, Classification, and Preservation. By W. J. LUCAS, B.A. Very fully Illustrated, with about 40 Coloured Plates, and numerous Black-and-White Engravings. *In cloth gilt, by subscription, 21s. nett. The price will be raised on publication to not less than 30s. nett.*

Bulb Culture, Popular. A Practical and Handy Guide to the Successful Cultivation of Bulbous Plants, both in the Open and under Glass. By W. D. DRURY. New Edition. Fully Illustrated. *In paper, price 1s., by post 1s. 2d.*

Bunkum Entertainments : A Collection of Original Laughable Skits on Conjuring, Physiognomy, Juggling, Performing Fleas, Waxworks, Panorama, Phrenology, Phonograph, Second Sight, Lightning Calculators, Ventriloquism, Spiritualism, &c., to which are added Humorous Sketches, Whimsical Recitals, and Drawing-room Comedies. By ROBERT GANTHONY Illustrated. *In cloth, price 2s. 6d., by post 2s. 9d.*

Butterflies, The Book of British : A Practical Manual for Collectors and Naturalists. Splendidly Illustrated throughout with very accurate Engravings of the Caterpillars, Chrysalids, and Butterflies, both upper and under -sides, from drawings by the Author or direct from Nature. By W. J. LUCAS, B.A. *In cloth gilt, price 3s. 6d., by post 3s. 9d.*

Butterfly and Moth Collecting : Where to Search, and What to Do. By G. E. SIMMS. Illustrated. *In paper, price 1s., by post 1s. 2d.*

Cabinet Making for Amateurs. Being clear Directions How to Construct many Useful Articles, such as Brackets, Sideboard, Tables, Cupboards, and other Furniture. Illustrated. *In cloth gilt, price 2s. 6d., by post 2s. 9d.*

Cactus Culture for Amateurs : Being Descriptions of the various Cactuses grown in this country ; with Full and Practical Instructions for their Successful Cultivation. By W. WATSON, Assistant Curator of the Royal Botanic Gardens, Kew. New Edition. Profusely Illustrated. *In cloth, gilt, price 5s. nett, by post 5s. 4d.*

Cage Birds, Diseases of : Their Causes, Symptoms, and Treatment. A Handbook for everyone who keeps a Bird. By DR. W. T. GREENE, F.Z.S. *In paper, price 1s., by post 1s. 2d.*

Cage Birds, Notes on. Second Series. Being Practical Hints on the Management of British and Foreign Cage Birds, Hybrids, and Canaries. By various Fanciers. Edited by DR. W. T. GREENE. *In cloth gilt, price 6s., by post 6s. 6d.*

Canary Book. The Breeding, Rearing, and Management of all Varieties of Canaries and Canary Mules, and all other matters connected with this Fancy. By ROBERT L. WALLACE. Third Edition. *In cloth gilt, price 5s., by post 5s. 4d. ; with COLOURED PLATES, 6s. 6d., by post 6s. 10d.*

General Management of Canaries. Cages and Cage-making, Breeding, Managing, Mule Breeding, Diseases and their Treatment, Moulting, Pests, &c. Illustrated. *In cloth gilt, price 2s. 6d., by post 2s. 9d.*

Exhibition Canaries. Full Particulars of all the different Varieties, their Points of Excellence, Preparing Birds for Exhibition, Formation and Management of Canary Societies and Exhibitions. Illustrated. *In cloth gilt, price 2s. 6d., by post 2s. 9d.*

Cane Basket Work: A Practical Manual on Weaving Useful and Fancy Baskets By ANNIE FIRTH. Illustrated. *In cloth gilt, price 1s. 6d., by post 1s. 8d.*

Card Conjuring: Being Tricks with Cards, and How to Perform Them. By PROF. ELLIS STANYON. Illustrated, and in Coloured Wrapper. *Price 1s., by post 1s. 2d.*

Card Tricks, Book of, for Drawing-room and Stage Entertainments by Amateurs; with an exposure of Tricks as practised by Card Sharpers and Swindlers. Numerous Illustrations. By PROF. R. KUNARD. *In illustrated wrapper, price 2s. 6d., by post 2s. 9d.*

Carnation Culture, for Amateurs. The Culture of Carnations and Picotees of all Classes in the Open Ground and in Pots. By B. C. RAVENSCROFT. Illustrated. *In paper, price 1s., by post 1s. 2d.*

Cats, Domestic or Fancy: A Practical Treatise on their Antiquity, Domestication, Varieties, Breeding, Management, Diseases and Remedies, Exhibition and Judging. By JOHN JENNINGS. Illustrated. *In cloth gilt, price 2s. 6d., by post 2s. 9d.*

Chrysanthemum Culture, for Amateurs and Professionals. Containing Full Directions for the Successful Cultivation of the Chrysanthemum for Exhibition and the Market. By B. C. RAVENSCROFT. New Edition. Illustrated. *In paper, price 1s., by post 1s. 2d.*

Chrysanthemum, The Show, and Its Cultivation. By C. SCOTT, of the Sheffield Chrysanthemum Society. *In paper, price 6d., by post 7d.*

Coins, a Guide to English Pattern, in Gold, Silver, Copper, and Pewter, from Edward I. to Victoria, with their Value. By the REV. G. F. CROWTHER, M.A. Illustrated. *In silver cloth, with gilt facsimiles of Coins, price 5s., by post 5s. 3d.*

Coins of Great Britain and Ireland, a Guide to the, in Gold, Silver, and Copper, from the Earliest Period to the Present Time, with their Value. By the late COLONEL W. STEWART THORBURN. Third Edition. Revised and Enlarged, by H. A. GRUEBER, F.S.A. Illustrated. *In cloth gilt, price 10s. 6d. nett, by post 10s. 10d.*

Cold Meat Cookery. A Handy Guide to making really tasty and much appreciated Dishes from Cold Meat. By MRS. J. E. DAVIDSON. *In paper, price 1s., by post 1s. 2d.*

Collie, The. Its History, Points, and Breeding. By HUGH DALZIEL. Illustrated with Coloured Frontispiece and Plates. *In paper, price 1s., by post 1s. 2d. ; cloth gilt, 2s., by post 2s 3d.*

Collie Stud Book. Edited by HUGH DALZIEL. *In cloth gilt, price 3s. 6d. each, by post 3s. 9d. each.*

> *Vol. I.,* containing Pedigrees of 1308 of the best-known Dogs, traced to their most remote known ancestors ; Show Record to Feb., 1890, &c.
>
> *Vol. II.* Pedigrees of 795 Dogs, Show Record, &c.
>
> *Vol. III.* Pedigrees of 786 Dogs, Show Record, &c.

Conjuring, Book of Modern. A Practical Guide to Drawing-room and Stage Magic for Amateurs. By PROFESSOR R. KUNARD. Illustrated. *In illustrated wrapper, price 2s. 6d., by post 2s. 9d.*

Conjuring for Amateurs. A Practical Handbook on How to Perform a Number of Amusing Tricks. By PROF. ELLIS STANYON. *In paper, price 1s., by post 1s. 2d.*

Cookery, The Encyclopædia of Practical. A complete Dictionary of all pertaining to the Art of Cookery and Table Service. Edited by THEO. FRANCIS GARRETT, assisted by eminent Chefs de Cuisine and Confectioners. Profusely Illustrated with Coloured Plates and Engravings by HAROLD FURNESS, GEO. CRUIKSHANK, W. MUNN ANDREW, and others. *In 2 vols., demy 4to., half morocco, cushion edges, price £3 3s.; carriage free, £3 5s.*

Cookery for Amateurs; or, French Dishes for English Homes of all Classes. Includes Simple Cookery, Middle-class Cookery, Superior Cookery, Cookery for Invalids, and Breakfast and Luncheon Cookery. By MADAME VALÉRIE. Second Edition. *In paper, price 1s., by post 1s. 2d.*

Cucumber Culture for Amateurs. Including also Melons, Vegetable Marrows and Gourds. Illustrated. By W. J. MAY. *In paper, price 1s., by post 1s. 2d.*

Cyclist's Route Map of England and Wales. Shows clearly all the Main, and most of the Cross, Roads, Railroads, and the Distances between the Chief Towns, as well as the Mileage from London. In addition to this, Routes of *Thirty of the Most Interesting Tours* are printed in red. Fourth Edition, thoroughly revised. The map is printed on specially prepared vellum paper, and is the fullest, handiest, and best up-to-date tourist's map in the market. *In cloth, price 1s., by post 1s. 2d.*

Dainties, English and Foreign, and How to Prepare Them. By MRS. DAVIDSON. *In paper, price 1s., by post 1s. 2d.*

Designing, Harmonic and Keyboard. Explaining a System whereby an endless Variety of Most Beautiful Designs suited to numberless Manufactures may be obtained by Unskilled Persons from any Printed Music. Illustrated by Numerous Explanatory Diagrams and Illustrative Examples. By C. H. WILKINSON. *Demy 4to, cloth gilt, price £3 3s. nett, by post £3 3s. 8d.*

Dogs, Breaking and Training: Being Concise Directions for the proper education of Dogs, both for the Field and for Companions. Second Edition. By "PATHFINDER." With Chapters by HUGH DALZIEL. Illustrated. *In cloth gilt, price 6s. 6d., by post 6s. 10d.*

Dogs, British, Ancient and Modern: Their Varieties, History, and Characteristics. By HUGH DALZIEL, assisted by Eminent Fanciers. Beautifully Illustrated with COLOURED PLATES and full-page Engravings of Dogs of the Day, with numerous smaller illustrations in the text. This is the fullest work on the various breeds of dogs kept in England. In three volumes, *demy 8vo, cloth gilt, price 10s. 6d. each, by post 11s. each.*

Vol. I. Dogs Used in Field Sports.

Vol. III. Practical Kennel Management: A Complete Treatise on all Matters relating to the Proper Management of Dogs whether kept for the Show Bench, for the Field, or for Companions.

Vol. II. is out of print, but Vols. I. and III. can still be had as above.

Dogs, Diseases of: Their Causes, Symptoms, and Treatment; Modes of Administering Medicines; Treatment in cases of Poisoning, &c. For the use of Amateurs. By HUGH DALZIEL. Fourth Edition. Entirely Re-written and brought up to Date. *In paper, price 1s., by post 1s. 2d.; in cloth gilt, 2s., by post 2s. 3d.*

Dog-Keeping, Popular: Being a Handy Guide to the General Management and Training of all Kinds of Dogs for Companions and Pets. By J. MAXTEE. Illustrated. *In paper, price 1s., by post 1s. 2d.*

Egg Dainties. How to Cook Eggs, One Hundred and Fifty Different Ways, English and Foreign. *In paper, price 1s., by post 1s. 2d.*

Engravings and their Value. Containing a Dictionary of all the Greatest Engravers and their Works. By J. H. SLATER. New Edition, Revised and brought up to date, with latest Prices at Auction. *In cloth gilt, price 15s. nett, by post, 15s. 5d.*

Entertainments, Amateur, for Charitable and other Objects: How to Organise and Work them with Profit and Success. By ROBERT GANTHONY. *In paper, price 1s., by post 1s. 2d.*

Fancy Work Series, Artistic. A Series of Illustrated Manuals on Artistic and Popular Fancy Work of various kinds. Each number is complete in itself, and issued at the uniform price of 6d., by post 7d. Now ready—(1) MACRAMÉ LACE (Second Edition); (2) PATCHWORK; (3) TATTING; (4) CREWEL WORK; (5) APPLIQUÉ; (6) FANCY NETTING.

Feathered Friends, Old and New. Being the Experience of many years' Observation of the Habits of British and Foreign Cage Birds. By DR. W. T. GREENE. Illustrated. *In cloth gilt, price 5s., by post 5s. 4d.*

Ferns, The Book of Choice: for the Garden, Conservatory, and Stove. Describing the best and most striking Ferns and Selaginellas, and giving explicit directions for their Cultivation, the formation of Rockeries, the arrangement of Ferneries, &c. By GEORGE SCHNEIDER. With numerous Coloured Plates and other Illustrations. *In 3 vols., large post 4to. Cloth gilt, price £3 3s. nett, by post £3 5s.*

Ferns, Choice British. Descriptive of the most beautiful Variations from the common forms, and their Culture. By C. T. DRUERY, F.L.S. Very accurate PLATES, and other Illustrations. *In cloth gilt, price 2s. 6d., by post 2s. 9d.*

Ferrets and Ferreting. Containing Instructions for the Breeding, Management, and Working of Ferrets. Second Edition, Re-written and greatly Enlarged. Illustrated. *In paper, price* 6d., *by post* 7d.

Fertility of Eggs Certificate. These are Forms of Guarantee given by the Sellers to the Buyers of Eggs for Hatching, undertaking to refund value of any unfertile eggs, or to replace them with good ones. Very valuable to sellers of eggs, as they induce purchases. *In books, with counterfoils. price* 6d., *by post* 7d.

Firework Making for Amateurs. A complete, accurate, and easily-understood work on Making Simple and High-class Fireworks. By DR. W. H. BROWNE, M.A. *In coloured wrapper, price* 2s 6d., *by post* 2s. 9d.

Fisherman, The Practical. Dealing with the Natural History, the Legendary Lore, the Capture of British Fresh-Water Fish, and Tackle and Tackle-making. By J. H. KEENE. *In cloth gilt, price* 7s. 6d., *by post* 7s. 10d.

Fish, Flesh, and Fowl. When in Season, How to Select, Cook, and Serve. By MARY BARRETT BROWN. *In paper, price* 1s., *by post* 1s. 3d.

Foreign Birds, Favourite, for Cages and Aviaries. How to Keep them in Health. By W. T. GREENE, M.A., M.D., F.Z.S., &c. Fully Illustrated. *In cloth gilt, price* 2s. 6d., *by post* 2s. 9d.

Fortune Telling by Cards. Describing and Illustrating the Methods usually followed by Persons Professing to Tell Fortunes by Cards. *Price* 1s., *by post* 1s. 2d.

Fox Terrier, The. Its History, Points, Breeding, Rearing, Preparing for Exhibition, and Coursing. By HUGH DALZIEL. Illustrated with Coloured Frontispiece and Plates. *In paper, price* 1s., *by post* 1s. 2d.; *cloth,* 2s., *by post* 2s. 3d.

Fox Terrier Stud Book. Edited by HUGH DALZIEL. *In cloth gilt, price* 3s. 6d. *each, by post* 3s. 9d. *each.*

> *Vol. I.,* containing Pedigrees of over 1400 of the best-known Dogs, traced to their most remote known ancestors.
> *Vol. II.* Pedigrees of 1544 Dogs, Show Record, &c.
> *Vol. III.* Pedigrees of 1214 Dogs, Show Record, &c.
> *Vol. IV.* Pedigrees of 1168 Dogs, Show Record, &c.
> *Vol. V.* Pedigrees of 1662 Dogs, Show Record, &c.

Fretwork and Marquetry. A Practical Manual of Instructions in the Art of Fret-cutting and Marquetry Work. By D. DENNING. Profusely Illustrated. *In cloth gilt, price* 2s. 6d., *by post* 2s. 9d.

Friesland Meres, A Cruise on the. By ERNEST R. SUFFLING. Illustrated. *In paper, price* 1s., *by post* 1s. 2d.

Fruit Culture for Amateurs. By S. T. WRIGHT. With Chapters on Insect and other Fruit Pests by W. D. DRURY. Second Edition. Illustrated. *In cloth gilt, price* 3s. 6d., *by post* 3s. 9d.

Game Preserving, Practical. Containing the fullest Directions for Rearing and Preserving both Winged and Ground Game, and Destroying Vermin; with other Information of Value to the Game Preserver. By W. CARNEGIE. Illustrated. *In cloth gilt, demy 8vo, price* 21s., *by post* 21s. 5d.

Games, the Book of a Hundred. By MARY WHITE. These Games are for Adults, and will be found extremely serviceable for Parlour Entertainment. They are Clearly Explained, are Ingenious, Clever, Amusing, and exceedingly Novel. *In stiff boards, price* 2s. 6d. *by post* 2s. 9d.

Gardening, Dictionary of. A Practical Encyclopædia of Horticulture, for Amateurs and Professionals. Illustrated with 2440 Engravings. Edited by G. NICHOLSON, Curator of the Royal Botanic Gardens, Kew; assisted by Prof. Trail, M.D., Rev. P. W. Myles, B.A., F.L.S., W. Watson. J. Garrett, and other Specialists. *In 4 vols., large post 4to. Cloth gilt, price* £3, *by post* £3 2s.

Gardening in Egypt. A Handbook of Gardening for Lower Egypt. With a Calendar of Work for the different Months of the Year. By WALTER DRAPER. *In cloth gilt, price* 3s. 6d., *by post* 3s. 9d.

Gardening, Home. A Manual for the Amateur, Containing Instructions for the Laying Out, Stocking, Cultivation, and Management of Small Gardens—Flower, Fruit, and Vegetable. By W. D. DRURY, F.R.H.S. Illustrated. *In paper, price* 1s., *by post* 1s. 2d.

Goat, Book of the. Containing Full Particulars of the Various Breeds of Goats, and their Profitable Management. With many Plates. By H. STEPHEN HOLMES PEGLER. Third Edition, with Engravings and Coloured Frontispiece. *In cloth gilt, price* 4s. 6d., *by post* 4s. 10d.

Goat-Keeping for Amateurs: Being the Practical Management of Goats for Milking Purposes. Abridged from "The Book of the Goat." Illustrated. *In paper, price* 1s., *by post* 1s. 2d.

Grape Growing for Amateurs. A Thoroughly Practical Book on Successful Vine Culture. By E. MOLYNEUX. Illustrated. *In paper, price 1s., by post 1s. 2d.*

Greenhouse Management for Amateurs. The Best Greenhouses and Frames, and How to Build and Heat them, Illustrated Descriptions of the most suitable Plants, with general and Special Cultural Directions, and all necessary information for the Guidance of the Amateur. By W. J. MAY. Second Edition, Revised and Enlarged. Magnificently Illustrated. *In cloth gilt, price 5s., by post 5s. 4d.*

Greyhound, The: Its History, Points, Breeding, Rearing, Training, and Running. By HUGH DALZIEL. With Coloured Frontispiece. *In cloth gilt, demy 8vo., price 2s. 6d., by post 2s. 9d.*

Guinea Pig, The, for Food, Fur, and Fancy. Its Varieties and its Management. By C. CUMBERLAND, F.Z.S. Illustrated. *In paper, price 1s., by post 1s. 2d. In cloth gilt, with coloured frontispiece, price 2s. 6d., by post 2s. 9d.*

Handwriting, Character Indicated by. With Illustrations in Support of the Theories advanced, taken from Autograph Letters, of Statesmen, Lawyers, Soldiers, Ecclesiastics, Authors, Poets, Musicians, Actors, and other persons. Second Edition. By R. BAUGHAN. *In cloth gilt, price 2s. 6d., by post 2s. 9d.*

Hardy Perennials and Old-fashioned Garden Flowers. Descriptions, alphabetically arranged, of the most desirable Plants for Borders, Rockeries, and Shrubberies, including Foliage as well as Flowering Plants. By J. WOOD. Profusely Illustrated. *In cloth gilt, price 3s. 6d., by post 3s. 9d.*

Hawk Moths, Book of British. A Popular and Practical Manual for all Lepidopterists. Copiously illustrated in black and white from the Author's own exquisite Drawings from Nature. By W. J. LUCAS, B.A. *In cloth gilt, price 3s. 6d., by post 3s. 9d.*

Home Medicine and Surgery: A Dictionary of Diseases and Accidents, and their proper Home Treatment. For Family Use. By W. J. MACKENZIE, M.D. Illustrated. *In cloth gilt, price 2s. 6d., by post 2s, 9d.*

Horse-Keeper, The Practical. By GEORGE FLEMING, C.B., LL.D., F.R.C.V.S., late Principal Veterinary Surgeon to the British Army, and Ex-President of the Royal College of Veterinary Surgeons. *In cloth gilt, price 3s. 6d., by post 3s. 10d.*

Horse-Keeping for Amateurs. A Practical Manual on the Management of Horses, for the guidance of those who keep one or two for their personal use. By FOX RUSSELL. *In paper, price 1s., by post 1s. 2d. ; cloth gilt 2s., by post 2s. 3d.*

Horses, Diseases of: Their Causes, Symptoms, and Treatment. For the use of Amateurs. By HUGH DALZIEL. *In paper, price 1s., by post 1s. 2d. ; cloth gilt 2s., by post 2s. 3d.*

Incubators and their Management. By J. H. SUTCLIFFE. New Edition, Revised and Enlarged. Illustrated. *In paper, price 1s., by post 1s. 2d.*

Inland Watering Places. A Description of the Spas of Great Britain and Ireland, their Mineral Waters, and their Medicinal Value, and the attractions which they offer to Invalids and other Visitors. Profusely illustrated. A Companion Volume to "Seaside Watering Places." *In cloth gilt, price 2s. 6d., by post 2s. 10d.*

Jack All Alone. Being a Collection of Descriptive Yachting Reminiscences. By FRANK COWPER, B.A., Author of "Sailing Tours." Illustrated. *In cloth gilt, price 3s. 6d., by post 3s. 10d.*

Journalism, Practical: How to Enter Thereon and Succeed. A book for all who think of "writing for the Press." By JOHN DAWSON. *In cloth gilt, price 2s. 6d., by post 2s. 9d.*

Lawn Tennis, Lessons in. A New Method of Study and Practise for Acquiring a Good and Sound Style of Play. With Exercises. By E. H. MILES. Illustrated. *In paper, price 1s., by post 1s. 2d.*

Laying Hens, How to Keep and to Rear Chickens in Large or Small Numbers, in Absolute Confinement, with Perfect Success. By MAJOR G. F. MORANT. *In paper, price 6d., by post 7d.*

Library Manual, The. A Guide to the Formation of a Library, and the Values of Rare and Standard Books. By J. H. SLATER, Barrister-at-Law. Third Edition. Revised and Greatly Enlarged. *In cloth gilt, price 7s. 6d. nett, by post 7s. 10d.*

Magic Lanterns, Modern. A Guide to the Management of the Optical Lantern, for the Use of Entertainers, Lecturers, Photographers, Teachers, and others. By R. CHILD BAYLEY. *In paper, price 1s., by post 1s. 2d.*

Marqueterie Painting for Amateurs. A Practical Handbook to Marqueterie, Wood-staining, and Kindred Arts. By ELIZA TURCK. Profusely Illustrated. *In paper, price 1s., by post 1s. 2d.*

Mice, Fancy: Their Varieties, Management, and Breeding. Third Edition, with additional matter and Illustrations. *In coloured wrapper representing different varieties, price 1s., by post 1s. 2d.*

Millinery, Handbook of. A Practical Manual of Instruction for Ladies. Illustrated. By MME. ROSÉE, Court Milliner, Principal of the School of Millinery. *In paper, price 1s., by post 1s. 2d.*

Model Yachts and Boats: Their Designing, Making, and Sailing. Illustrated with 118 Designs and Working Diagrams. By J. DU V. GROSVENOR. *In leatherette, price 5s., by post 5s. 3d.*

Monkeys, Pet, and How to Manage Them. Illustrated. By ARTHUR PATTERSON. *In cloth gilt, price 2s. 6d., by post 2s. 9d.*

Mountaineering, Welsh. A Complete and Handy Guide to all the Best Roads and Bye-Paths by which the Tourist should Ascend the Welsh Mountains. By A. W. PERRY. With numerous Maps. *In cloth gilt, price 2s. 6d., by post 2s. 9d.*

Mushroom Culture for Amateurs. With Full Directions for Successful Growth in Houses, Sheds, Cellars, and Pots, on Shelves, and Out of Doors. By W. J. MAY. Illustrated. *In paper, price 1s., by post 1s. 2d.*

Natural History Sketches among the Carnivora—Wild and Domesticated; with Observations on their Habits and Mental Faculties. By ARTHUR NICOLS, F.G.S., F.R.G.S. Illustrated. *In cloth gilt, price 2s. 6d., by post 2s. 9d.*

Naturalist's Directory, The, for 1899 (Fifth year of issue). Invaluable to all Students and Collectors. *In paper, price 1s., by post 1s.*

Needlework, Dictionary of. An Encyclopædia of Artistic, Plain, and Fancy Needlework; Plain, practical, complete, and magnificently Illustrated. By S. F. A. CAULFEILD and B. C. SAWARD. *In demy 4to, 528pp, 829 Illustrations, extra cloth gilt, plain edges, cushioned bevelled boards, price 21s. nett, by post 21s. 9d.; with COLOURED PLATES, elegant satin brocade cloth binding, and coloured edges, 31s. 6d. nett, by post 32s.*

Orchids: Their Culture and Management, with Descriptions of all the Kinds in General Cultivation. Illustrated by Coloured Plates and Engravings. By W. WATSON, Assistant-Curator, Royal Botanic Gardens, Kew; Assisted by W. BEAN, Foreman, Royal Gardens, Kew. Second Edition, Revised and with Extra Plates. *In cloth gilt and gilt edges, price £1 1s. nett, by post £1 1s. 6d.*

Painters and Their Works. A Work of the Greatest Value to Collector and such as are interested in the Art, as it gives, besides Biographical Sketches of all the Artists of Repute (not now living) from the 13th Century to the present date, the Market Value of the Principal Works Painted by Them, with Full Descriptions of Same. *In 3 vols., cloth gilt, price 37s. 6d. nett, by post 38s. 3d.*

Painting, Decorative. A practical Handbook on Painting and Etching upon Textiles, Pottery, Porcelain, Paper, Vellum, Leather, Glass, Wood, Stone, Metals, and Plaster. for the Decoration of our Homes. By B. C. SAWARD. *In cloth gilt, price 3s. 6d., by post 3s. 9d.*

Palmistry, Life Studies in. The Hands of Notable Persons read according to the practice of Modern Palmistry. By INA OXENFORD. Illustrated with Full-page Plates. *In 4to, cloth gilt, price 5s., by post 5s. 4d.*

Parcel Post Dispatch Book (registered). An invaluable book for all who send parcels by post. Provides Address Labels, Certificate of Posting, and Record of Parcels Dispatched. By the use of this book parcels are insured against loss or damage to the extent of £2. Authorised by the Post Office. *Price 1s., by post 1s. 2d., for 100 parcels; larger sizes if required.*

Parrakeets, Popular. How to Keep and Breed Them. By DR. W. T. GREENE, M.A., F.Z.S., &c. *In paper, price 1s., by post 1s. 2d.*

Parrot, The Grey, and How to Treat it. By W. T. GREENE, M.D., M.A., F.Z.S., &c. *In paper, price 1s., by post 1s. 2d.*

Parrots, the Speaking. The Art of Keeping and Breeding the principal Talking Parrots in Confinement. By DR. KARL RUSS. Illustrated with COLOURED PLATES and Engravings. *In cloth gilt, price 5s., by post 5s. 4d.*

Patience, Games of, for one or more Players. How to Play 142 different Games of Patience. By M. WHITMORE JONES. Illustrated. Series I., 39 games; Series II., 34 games; Series III., 33 games; Series IV., 37 games. Each, in paper, 1s., by post 1s. 2d. *The four bound together in cloth gilt, price 5s., by post 5s. 4d.*

Patience Cards, for Games of. Two dainty Packs (2¼in. by 2in.) for playing the Various Games of Patience, in Case. They are of the best make and finish, and of a very pretty and convenient size. *Price 2s. 6d., by post 2s. 9d.*

Pedigree Record, The. Being Part I. of "The Breeders and Exhibitors Record," for the Registration of Particulars concerning Pedigrees of Stock of every Description. By W. K. TAUNTON *In cloth gilt, price 2s. 6d., by post 2s. 9d.*

Perspective, The Essentials of. With numerous Illustrations drawn by the Author. By L. W. MILLER, Principal of the School of Industrial Art of the Pennsylvania Museum, Philadelphia. *Price 6s. 6d., by post 6s. 10d.*

Pheasant-Keeping for Amateurs. A Practical Handbook on the Breeding, Rearing, and General Management of Fancy Pheasants in Confinement. By GEO. HORNE. Fully Illustrated. *In cloth gilt, price 3s. 6d., by post 3s. 9d.*

Photographic Printing Processes, Popular. A Practical Guide to Printing with Gelatino-Chloride, Artigue, Platinotype, Carbon, Bromide, Collodio-Chloride, Bichromated Gum, and other Sensitised Papers. By H. MACLEAN, F.R.P.S. Illustrated. *In cloth gilt, price 2s. 6d., by post 2s. 10d.*

Photography (Modern) for Amateurs. Fourth Edition. Revised and Enlarged. By J. EATON FEARN. *In paper, price 1s., by post 1s. 2d.*

Pianofortes, Tuning and Repairing. The Amateur's Guide to the Practical Management of a Piano without the intervention of a Professional. By CHARLES BABBINGTON. *In paper, price 6d., by post 6½d.*

Picture-Frame Making for Amateurs. Being Practical Instructions in the Making of various kinds of Frames for Paintings, Drawings, Photographs, and Engravings. By the REV. J. LUKIN. Illustrated. *In paper, price 1s., by post 1s 2d.*

Pig, Book of the. The Selection, Breeding, Feeding, and Management of the Pig; the Treatment of its Diseases; the Curing and Preserving of Hams, Bacon, and other Pork Foods; and other information appertaining to Pork Farming. By PROFESSOR JAMES LONG. Fully Illustrated with Portraits of Prize Pigs, Plans of Model Piggeries, &c. *In cloth gilt, price 10s. 6d., by post 10s. 11d.*

Pig-Keeping, Practical: A Manual for Amateurs, based on personal Experience in Breeding, Feeding, and Fattening; also in Buying and Selling Pigs at Market Prices. By R. D. GARRATT. *In paper, price 1s., by post 1s. 2d.*

Pigeons, Fancy. Containing full Directions for the Breeding and Management of Fancy Pigeons, and Descriptions of every known Variety, together with all other information of interest or use to Pigeon Fanciers. Third Edition. 18 COLOURED PLATES, and 22 other full-page Illustrations. By J. C. LYELL. *In cloth gilt, price 10s. 6d., by post 10s. 10d.*

Pigeon-Keeping for Amateurs. A Complete Guide to the Amateur Breeder of Domestic and Fancy Pigeons. By J. C. LYELL. Illustrated. *In cloth gilt, price 2s. 6d., by post 2s. 9d.; in paper, price 1s., by post 1s. 2d.*

Polishes and Stains for Wood: A Complete Guide to Polishing Woodwork, with Directions for Staining, and Full Information for Making the Stains, Polishes, &c., in the simplest and most satisfactory manner. By DAVID DENNING. *In paper, 1s., by post 1s. 2d.*

Pool, Games of. Describing Various English and American Pool Games, and giving the Rules in full. Illustrated *In paper, price 1s., by post 1s. 2d.*

Portraiture, Home, for Amateur Photographers. Being the result of many years' incessant work in the production of Portraits "at home." By RICHARD PENLAKE. Fully Illustrated. *In cloth gilt, price 2s. 6d., by post 2s. 9d.*

Postage Stamps, and their Collection. A Practical Handbook for Collectors of Postal Stamps, Envelopes, Wrappers, and Cards. By OLIVER FIRTH, Member of the Philatelic Societies of London, Leeds, and Bradford. Profusely Illustrated. *In cloth gilt, price 3s. 6d., by post 3s. 10d.*

Postage Stamps of Europe, The Adhesive: A Practical Guide to their Collection, Identification, and Classification. Especially designed for the use of those commencing the Study. By W. A. S. WESTOBY. Beautifully Illustrated. *In paper Parts, 1s. each, by post 1s. 2d. Vol. I., cloth gilt, price 7s. 6d., by post 8s.*

Postmarks, History of British. With 350 Illustrations and a List of Numbers used in Obliterations. By J. H. DANIELS. *In cloth gilt, price 2s. 6d., by post 2s. 9d.*

Pottery and Porcelain, English. A Guide for Collectors. Handsomely Illustrated with Engravings of Specimen Pieces and the Marks used by the different Makers. New Edition, Revised and Enlarged. By the REV. E. A. DOWNMAN. *In cloth gilt, price 5s., by post 5s. 3d.*

Poultry-Farming, Profitable. Describing in Detail the Methods that Give the Best Results, and pointing out the Mistakes to be Avoided. By J. H. Sutcliffe. Illustrated. *In paper, price* 1s., *by post* 1s. 2d.

Poultry-Keeping, Popular. A Practical and Complete Guide to Breeding and Keeping Poultry for Eggs or for the Table. By F. A. Mackenzie. Illustrated. *In paper, price* 1s., *by post* 1s. 2d.

Poultry for Prizes and Profit. Contains: Breeding Poultry for Prizes, Exhibition Poultry and Management of the Poultry Yard. Handsomely Illustrated. Second Edition. By Prof. James Long. *In cloth gilt, price* 2s. 6d., *by post* 2s. 10d.

Rabbit, Book of The. A Complete Work on Breeding and Rearing all Varieties of Fancy Rabbits, giving their History, Variations, Uses, Points, Selection, Mating, Management, &c., &c. SECOND EDITION. Edited by Kempster W. Knight. Illustrated with Coloured and other Plates. *In cloth gilt, price* 10s. 6d., *by post* 10s. 11d.

Rabbits, Diseases of: Their Causes, Symptoms, and Cure. With a Chapter on The Diseases of Cavies. Reprinted from "The Book of the Rabbit" and "The Guinea Pig for Food, Fur, and Fancy." *In paper, price* 1s., *by post* 1s. 2d.

Rabbits for Prizes and Profit. The Proper Management of Fancy Rabbits in Health and Disease, for Pets or the Market, and Descriptions of every known Variety, with Instructions for Breeding Good Specimens. By Charles Rayson. Illustrated. *In cloth gilt, price* 2s. 6d., *by post* 2s. 9d. Also in Sections, as follows:

> *General Management of Rabbits.* Including Hutches, Breeding, Feeding, Diseases and their Treatment, Rabbit Courts, &c. Fully Illustrated. *In paper, price* 1s., *by post* 1s. 2d.

> *Exhibition Rabbits.* Being descriptions of all Varieties of Fancy Rabbits, their Points of Excellence, and how to obtain them. Illustrated. *In paper, price* 1s., *by post* 1s. 2d.

Road Charts (Registered). For Army Men, Volunteers, Cyclists, and other Road Users. By S. W. H. Dixon and A. B. H. Clerke. No. 1.—London to Brighton. *Price* 2d., *by post* 2½d.

Roses for Amateurs. A Practical Guide to the Selection and Cultivation of the best Roses. Illustrated. By the Rev. J. Honywood D'Ombrain, Hon. Sec. Nat. Rose Soc. *In paper, price* 1s., *by post* 1s. 2d.

Sailing Guide to the Solent and Poole Harbour, with Practical Hints as to Living and Cooking on, and Working a Small Yacht. By Lieut.-Col. T. G. Cuthell. Illustrated with Coloured Charts. *In cloth gilt, price* 2s. 6d., *by post* 2s. 9d.

Sailing Tours. The Yachtman's Guide to the Cruising Waters of the English and Adjacent Coasts. With Descriptions of every Creek, Harbour, and Roadstead on the Course. With numerous Charts printed in Colours, showing Deep water, Shoals, and Sands exposed at low water, with sounding. By Frank Cowper, B.A. *In Crown 8vo., cloth gilt.*

> *Vol. I.,* the Coasts of Essex and Suffolk, from the Thames to Aldborough. Six Charts. *Price* 5s., *by post* 5s. 3d.

> *Vol. II.* The South Coast, from the Thames to the Scilly Islands, twenty-five Charts. New and Revised Edition. *Price* 7s. 6d., *by post* 7s. 10d.

> *Vol. III.* The Coast of Brittany, from L'Aberwrach to St. Nazaire, and an Account of the Loire. Twelve Charts. *Price* 7s. 6d., *by post* 7s. 10d.

> *Vol. IV.* The West Coast, from Land's End to Mull of Galloway, including the East Coast of Ireland. Thirty Charts. *Price* 10s. 6d., *by post* 10s. 10d.

> *Vol. V.* The Coasts of Scotland and the N.E. of England down to Aldborough. Forty Charts. *Price* 10s. 6d., *by post* 10s. 10d.

St. Bernard, The. Its History, Points, Breeding, and Rearing. By Hugh Dalziel. Illustrated with Coloured Frontispiece and Plates. *In cloth gilt, price* 2s. 6d., *by post* 2s. 9d.

St. Bernard Stud Book. Edited by Hugh Dalziel. *In cloth gilt, price* 3s. 6d. each, *by post* 3s. 9d. each.

> *Vol. I.* Pedigrees of 1278 of the best known Dogs traced to their most remote known ancestors, Show Record, &c.

> *Vol. II.* Pedigrees of 564 Dogs, Show Record, &c.

Sea-Fishing for Amateurs. Practical Instructions to Visitors at Sea-side Places for Catching Sea-Fish from Pier-heads, Shore, or Boats, principally by means of Hand Lines, with a very useful List of Fishing Stations, the Fish to be caught there, and the Best Seasons. By FRANK HUDSON. Illustrated. *In paper, price 1s., by post 1s. 2d.*

Sea-Life, Realities of. Describing the Duties, Prospects, and Pleasures of a Young Sailor in the Mercantile Marine. By H. E. ACRAMAN COATE. With a Preface by J. R. DIGGLE, M.A., M.L.S.B. *In cloth gilt, price 3s. 6d., by post 3s. 10d.*

Seaside Watering Places. A Description of the Holiday Resorts on the Coasts of England and Wales, the Channel Islands, and the Isle of Man, giving full particulars of them and their attractions, and all information likely to assist persons in selecting places in which to spend their Holidays according to their individual tastes. Illustrated. Twenty-third Year of Issue. *In cloth gilt, price 2s. 6d., by post 2s. 10d.*

Sea Terms, a Dictionary of. For the use of Yachtsmen, Amateur Boatmen, and Beginners. By A. ANSTED. Fully Illustrated. *In cloth gilt, price 7s. 6d. nett, by post 7s. 11d.*

Shadow Entertainments, and How to Work them : being Something about Shadows, and the way to make them Profitable and Funny. By A. PATTERSON. Illustrated. *In paper, price 1s., by post 1s. 2d.*

Shave, An Easy: The Mysteries, Secrets, and Whole Art of, laid bare. Edited by JOSEPH MORTON. *Price 1s., by post 1s. 2d.*

Sheep Raising and Shepherding. A Handbook of Sheep Farming. By W. J. MALDEN, Principal of the Agricultural College, Uckfield. Illustrated. *Cloth gilt, price 3s. 6d., by post 3s. 9d.*

Sheet Metal, Working in: Being Practical Instructions for Making and Mending Small Articles in Tin, Copper, Iron, Zinc, and Brass. By the Rev. J. LUKIN, B.A. Illustrated. Third Edition. *In paper, price 1s., by post 1s. 1d.*

Show Record, The. Being Part III. of "The Breeders' and Exhibitors' Record," for the Registration of Particulars concerning the Exhibition of Pedigree stock of every Description. By W. K. TAUNTON. *In cloth gilt, price 2s. 6d., by post 2s. 9d.*

Skating Cards: An Easy Method of Learning Figure Skating, as the Cards can be used on the Ice. *In cloth case, price 2s. 6d., by post 2s. 9d.; leather, price 3s. 6d., by post 3s. 9d.* A cheap form is issued printed on paper and made up as a small book, *price 1s., by post 1s. 1d.*

Sleight of Hand. A Practical Manual of Legerdemain for Amateurs and Others. New Edition, Revised and Enlarged. Profusely Illustrated. By E. SACHS. *In cloth gilt, price 6s. 6d., by post 6s. 10d.*

Solo Whist. A Practical Manual both for Beginners and Advanced Students, with Amended Exhaustive Code of Laws. By C. J. MELROSE. *In cloth gilt, price 3s. 6d., by post 3s. 10d.*

Sporting Books, Illustrated. A Descriptive Survey of a Collection of English Illustrated Works of a Sporting and Racy Character, with an Appendix of Prints relating to Sports of the Field. The whole valued by reference to Average Auction Prices. By J. H. SLATER, Author of "Library Manual," "Engravings and Their Value," &c. *In cloth gilt, price 7s. 6d. nett, by post 7s. 10d.*

Stud Record, The. Being Part II. of "The Breeders' and Exhibitors' Record," for the Registration of Particulars concerning Pedigree Stock of every Description. By W. K. TAUNTON. *In cloth gilt, price 2s. 6d., by post 2s. 9d.*

Taxidermy, Practical. A Manual of Instruction to the Amateur in Collecting, Preserving, and Setting-up Natural History Specimens of all kinds. With Examples and Working Diagrams. By MONTAGU BROWNE, F.Z.S., Curator of Leicester Museum. Second Edition. *In cloth gilt, price 7s. 6d., by post 7s. 10d.*

Thames Guide Book. From Lechlade to Richmond. For Boating Men, Anglers, Picnic Parties, and all Pleasure-seekers on the River. Arranged on an entirely new plan. Second Edition, profusely Illustrated. *In paper, price 1s., by post 1s. 2d.*

Tomato and Fruit Growing as an Industry for Women. Lectures given at the Forestry Exhibition, Earl's Court, during July and August, 1893. By GRACE HARRIMAN, Practical Fruit Grower and County Council Lecturer. *In paper, price 1s., by post 1s. 1d.*

Tomato Culture for Amateurs. A Practical and very Complete Manual on the subject. By B. C. RAVENSCROFT. Illustrated. *In paper, price 1s., by post 1s. 1d.*

Trapping, Practical: Being some Papers on Traps and Trapping for Vermin, with a Chapter on General Bird Trapping and Snaring. By W. CARNEGIE. In *paper, price 1s., by post 1s. 2d.*

Turning Lathes. A Manual for Technical Schools and Apprentices. A Guide to Turning, Screw-cutting, Metal-spinning, &c. Edited by JAMES LUKIN, B.A. Third Edition. With 194 Illustrations. *In cloth gilt, price 3s., by post 3d. 3s.*

Vamp, How to. A Practical Guide to the Accompaniment of Songs by the Unskilled Musician. With Examples. *In paper, price 9d., by post 10d.*

Vegetable Culture for Amateurs. Containing Concise Directions for the Cultivation of Vegetables in small Gardens so as to insure Good Crops. With Lists of the Best Varieties of each Sort. By W. J. MAY Illustrated. *In paper, price 1s., by post 1s. 2d.*

Ventriloquism, Practical. A thoroughly reliable Guide to the Art of Voice Throwing and Vocal Mimicry, Vocal Instrumentation, Ventriloquial Figures, Entertaining, &c. By ROBERT GANTHONY. Numerous Illustrations. *In cloth gilt, price 2s. 6d., by post 2s. 9d.*

Violins (Old) and their Makers: Including some References to those of Modern Times. By JAMES M. FLEMING. Illustrated with Facsimiles of Tickets, Sound-Holes, &c. *In cloth gilt, price 6s. 6d. nett, by post 6s. 10d.*

Violin School, Practical, for Home Students. Instructions and Exercises in Violin Playing, for the use of Amateurs, Self-learners, Teachers, and others. With a Supplement on "Easy Legato Studies for the Violin." By J. M. FLEMING. *Demy 4to, cloth gilt, price 9s. 6d., by post 10s. 2d.* Without Supplement, *price 7s. 6d., by post 8s.*

Vivarium, The. Being a Full Description of the most Interesting Snakes, Lizards, and other Reptiles, and How to Keep Them Satisfactorily in Confinement. By REV. G. C. BATEMAN. Beautifully Illustrated. *In cloth gilt, price 7s. 6d. nett, by post 8s.*

War Medals and Decorations. A Manual for Collectors. with some account of Civil Rewards for Valour. By D. HASTINGS IRWIN. Revised and Enlarged Edition. Beautifully Illustrated. *In cloth gilt, price 12s. 6d. nett, by post 12s. 10d.*

Whippet and Race-Dog, The: How to Breed, Rear, Train, Race, and Exhibit the Whippet, the Management of Race Meetings, and Original Plans of Courses. By FREEMAN LLOYD. *In cloth gilt, price 3s. 6d., by post 3s. 10d.*

Whist, Scientific: Its Whys and Wherefores. Wherein all Arbitrary Dicta of Authority are eliminated, the Reader being taken step by step through the Reasoning Operations upon which the Rules of Play are based. By C. J. MELROSE. With Illustrative Hands printed in Colour. *In cloth gilt, price 6s., by post 6s. 6d.*

Wild Birds, Cries and Call Notes of, Described at Length, and in many instances Illustrated by Musical Notation. *In paper, price 1s., by post 1s. 2d.*

Wildfowling, Practical: A Book on Wildfowl and Wildfowl Shooting. By HY. SHARP. The result of 25 years experience Wildfowl Shooting under all sorts of conditions of locality as well as circumstances. Profusely Illustrated. *Demy 8vo, cloth gilt, price 12s. 6d. nett, by post 12s. 10d.*

Wild Sports in Ireland. Being Picturesque and Entertaining Descriptions of several visits paid to Ireland, with Practical Hints likely to be of service to the Angler, Wildfowler, and Yachtsman. By JOHN BICKERDYKE, Author of "The Book of the All-Round Angler," &c. Beautifully Illustrated from Photographs taken by the Author. *In cloth gilt, price 6s., by post 6s. 4d.*

Window Ticket Writing. Containing full Instructions on the Method of Mixing and Using the Various Inks, &c., required, Hints on Stencilling as applied to Ticket Writing, together with Lessons on Glass Writing, Japanning on Tin, &c. Especially written for the use of Learners and Shop Assistants. By WM. C. SCOTT. *In paper, price 1s., by post 1s. 2d.*

* 9 7 8 3 7 4 2 8 6 0 8 1 1 *